IPS

Research Guide to Current Military and Strategic Affairs

William M. Arkin

The Institute for Policy Studies is a nonpartisan research institute. The views expressed in this study are solely those of the author.

All rights reserved. Reprinting by permission.

© 1981 William M. Arkin

Published by the Institute for Policy Studies.

Copies of this book are available from the Institute for Policy Studies, 1901 Q Street, N.W., Washington, D.C. 20009 or the Transnational Institute, Paulus Potterstraat 20, 1071 DA, Amsterdam, Holland.

First Printing, 1981

ISBN 0-89758-025-7 paper
ISBN 0-89758-032-X cloth

ABOUT THE AUTHOR

William M. Arkin has conducted research in the areas of United States military organization and programs, nuclear weapons and comparative military analysis. He is a former senior staff analyst at the Center for Defense Information in Washington, D.C., and a former Army intelligence analyst who served in Berlin for over three years. He is currently a graduate student in National Security Studies at Georgetown University.

TABLE OF CONTENTS

LIST OF TABLES ix

ABBREVIATIONS USED IN THIS GUIDE xii

I. INTRODUCTION 1

II. GENERAL INFORMATION SOURCES 3
  A. Reference Tools 3
    1. Guides to Research 3
    2. Bibliographies 5
    3. Abstracts and Indexes 8
    4. Dictionaries 12
  B. Data Sources 14
    1. General Data 14
    2. Armed Forces 16
    3. Military Expenditures 17
    4. The Arms Trade 18
  C. Descriptive Sources 19
    1. Events 20
    2. Personalities 21
    3. Reviews 22

III. U.S. GOVERNMENT DOCUMENTS AND INFORMATION 27
  A. Introduction 27
    1. Guides to Research 28
    2. Abstracts and Indexes 28
    3. Government Organization 29
    4. The Freedom of Information Act 31
  B. Congressional Information 33
    1. The Congressional Documents System 34
    2. Defense and Foreign Policy Sources 37
    3. Congressional Agencies 39
  C. Executive Branch Information 42
    1. Presidential Information 42
    2. The Central Intelligence Agency 43
    3. The Department of State 45
    4. The Department of Defense 48
      (a) DOD Information 50
      (b) Air Force Information 53
      (c) Army Information 55

      (d) Navy Information *59*
      (e) Marine Corps Information *62*
      (f) Technical Information *63*

IV. THE U.S. MILITARY *67*
A. Department of Defense Organization and Background *67*
    1. *DOD and DOD Agencies* *68*
    2. *The Air Force* *70*
    3. *The Army* *73*
    4. *The Navy* *77*
    5. *The Marine Corps* *81*
    6. *The Reserves* *82*

B. Defense Policy and Posture *85*
    1. *U.S. National Security Policy* *85*
      (a) Background *86*
      (b) Current Defense Posture *87*
      (c) U.S. Foreign Policy *88*
      (d) The U.S.-Soviet Military Balance *90*
    2. *Air Force Current Posture* *91*
    3. *Army Current Posture* *93*
    4. *Navy Current Posture* *94*
    5. *Marine Corps Current Posture* *95*
    6. *Reserve Forces Current Posture* *96*

C. The Defense Budget *97*
    1. *The Budget Process and Budget Formulation* *97*
    2. *The Annual Cycle* *98*

D. The Military-Industrial Complex *104*
    1. *Researching the Complex and Companies* *104*
    2. *Defense Industries and Contractors* *106*
    3. *Contracting and Procurement Policy* *109*
    4. *Current Procurement* *110*
    5. *Research & Development* *111*

E. Military Personnel *114*
    1. *Policy, Costs and Compensation* *114*
    2. *Manpower Strength Figures* *116*
    3. *The All-Volunteer Armed Forces and the Draft* *117*
    4. *Women in the Military* *118*

F. The Local Impact of Military Presence and Spending *119*
    1. *Factors Affecting the Community* *120*
    2. *Sources of Comparative Data* *121*
    3. *Military Bases* *121*

G. Overseas Bases, Commitments and Aid *124*
    1. *Commitments* *125*

    2. Overseas Bases 127
    3. Regional Presence and Relations 128
    4. Aid and Military Assistance 129
       (a) Policy and Implementation 129
       (b) Current Programs 131

V. WORLDWIDE MILITARY AND STRATEGIC AFFAIRS 133

A. Armed Forces and Weapons 133
    1. Information Services 134
    2. Weapons Series 137
    3. Air Forces 141
       (a) Aircraft and Weapons 143
    4. Ground Forces 144
       (a) Weapons 145
    5. Naval Forces 146
       (a) Ships and Weapons 147
    6. Nuclear Weapons 148
    7. Missiles 149
    8. Military Communications and Electronic Warfare 150

B. Regional and Country Defense Issues 151
    1. African Defense Issues 152
       (a) Reference Tools 153
       (b) Current Issues and Events 154
    2. Asian Defense Issues 156
       (a) Reference Tools 156
       (b) Current Issues and Events 157
       (c) Australian Defense Issues 159
       (d) Indian Defense Issues 160
       (e) Japanese Defense Issues 161
       (f) Korean Defense Issues 162
       (g) Other Country Defense Issues 163
    3. Chinese Defense Issues 164
       (a) Reference Tools 164
       (b) Current Issues and Events 164
       (c) Chinese Foreign Policy 165
       (d) The Chinese Military 165
    4. Latin American Defense Issues 166
       (a) Reference Tools 166
       (b) Current Issues and Events 167
    5. Middle East Defense Issues 168
       (a) Reference Tools 168
       (b) Current Issues and Events 169
       (c) The Arab-Israeli Conflict 169
       (d) Persian Gulf Security 170

- **C. The Soviet Union, Eastern Europe and the Warsaw Pact** 171
  - 1. Reference Tools 171
  - 2. Current Issues and Events 172
  - 3. Soviet Foreign Policy and Relations 174
  - 4. Soviet and East European Military Forces 175
    - (a) Background 175
    - (b) Ground Forces .... .176
    - (c) Air Forces 177
    - (d) Naval Forces 177
  - 5. Soviet Military Doctrine and Strategy 178
  - 6. Soviet Military Spending 178
- **D. Western Europe and NATO** 181
  - 1. Reference Tools 181
  - 2. Current Issues and Events 181
  - 3. The United Kingdom 182
  - 4. Northern European Defense Issues 184
  - 5. Central European Defense Issues 185
  - 6. Southern European Defense Issues 186
  - 7. NATO 187
  - 8. Canada 189
- **E. Arms Control and Disarmament** 190
  - 1. Reference Tools 191
  - 2. Background 192
- **F. International Organizations and International Law** 194
  - 1. International Law 194
  - 2. International Organizations 195
  - 3. The United Nations 196

**APPENDIX A:
MILITARY AND STRATEGIC SERIALS
AND PERIODICALS** 197

**APPENDIX B:
ORGANIZATIONS** 228

LIST OF TABLES

TABLE 1
General International Relations Journals 25

TABLE 2
Strategic and Military Journals 26

TABLE 3
Freedom of Information Offices of the National
Security Community 32

TABLE 4
Congressional Committees and Subcommittees
in the National Security Field 35

TABLE 5
Annual Congressional Hearings in National
Security Affairs 38

TABLE 6
Joint Publications Research Service Translation Series 44

TABLE 7
Periodicals Published by the Air Force 56

TABLE 8
Periodicals Published by the Army 57

TABLE 9
Army Regulatory Documents and Publications
(short and long titles) 58

TABLE 10
Periodicals Published by the Navy 60

TABLE 11
Navy Regulatory Documents and Publications
(short and long titles) 61

TABLE 12
Higher Military Educational Institutions of the DOD 65

TABLE 13
Department of Defense Agencies
and Unified Commands 69

TABLE 14
Air Force Major Commands and Separate
Operating Agencies 71

TABLE 15
Basic Documents on Air Force Doctrine 73

TABLE 16
Army Major Commands 75

TABLE 17
Basic Documents on Army Doctrine 76

TABLE 18
Navy Major Commands and Operating Forces 78

TABLE 19
Basic Documents on Naval Doctrine 81

TABLE 20
Basic Documents on Marine Corps Doctrine 83

TABLE 21
Newsletters on Current Defense Posture and Programs 89

TABLE 22
Defense Budget Cycle 99

TABLE 23
Appropriation Titles within the Defense Budget 101

TABLE 24
Defense Associations and Publications 109

TABLE 25
Information Services and Products of DMS, Inc. 135

TABLE 26
Military Newsletters 138

TABLE 27
Jane's Pocket Books 139

TABLE 28
Periodicals Dealing with Air Forces and Weapons 143

TABLE 29
Periodicals Dealing with Ground Forces and Weapons 145

TABLE 30
Periodicals Dealing with Naval Forces and Ships
and Weapons 147

TABLE 31
African Journals and Magazines 154

TABLE 32
Asian Journals and Magazines 158

TABLE 33
CIA Translation Services on the U.S.S.R.
and Eastern Europe 173

TABLE 34
Soviet Military Thought Series 179

TABLE 35
DIA Publications on the Soviet Union 180

TABLE 36
British Military Periodicals 184

TABLE 37
Subject Listings of Periodicals 197

TABLE 38
Serials and Monographs in Military
and Strategic Affairs 198

ABBREVIATIONS USED IN THIS GUIDE

ACDA	Arms Control and Disarmanent Agency
ADIU	Armament and Disarmament Information Unit (Univ. of Sussex)
AEI	American Enterprise Institute for Public Policy Research
AFB	Air Force Base
AFSC	American Friends Service Committee
AID	Agency for International Development
aka	also known as
ASD	Assistant Secretary of Defense
CBO	Congressional Budget Office
CBR	Chemical, Biological, and Radiological Warfare
CCH	Commerce Clearing House
CIA	Central Intelligence Agency
CIS	Congressional Information Service
CNA	Center for Naval Analysis
CRS	Congressional Research Service
CSIS	Center for Strategic and International Studies (Georgetown Univ.)
CSU	California State University
DA	Department of the Army
DCAA	Defense Contract Audit Agency
DIA	Defense Intelligence Agency
DIOR	Directorate of Information and Operations Reports
DMS	Defense Marketing Service, Inc.
DOD	Department of Defense
DSAA	Defense Security Assistance Agency
FAS	Federation of American Scientists
FBIS	Foreign Broadcast Information Service
FCNL	Friends Committee on National Legislation
FOIA	Freedom of Information Act
FY	Fiscal Year
GAO	General Accounting Office
GPO	Government Printing Office
IDSA	Institute for Defence Studies and Analysis (India)

IISS	International Institute for Strategic Studies (London)
JAG	Judge Advocate General
JCS	Joint Chiefs of Staff
JPRS	Joint Publications Research Service
LC	Library of Congress
MAGTF	Marine Air Ground Task Force
MIT	Massachusetts Institute of Technology
MOD	Ministry of Defense
NACLA	North American Congress on Latin America
NATO	North Atlantic Treaty Organization
NBC	Nuclear, Biological and Chemical Warfare
NDU	National Defense University
NRL	Naval Research Laboratory
NSF	National Science Foundation
NTIS	National Technical Information Service
OASD (MRAL)	Office of the Assistant Secretary of Defense (Manpower, Reserve Affairs, and Logistics)
OASD (PA)	Office of the Assistant Secretary of Defense (Public Affairs)
OMB	Office of Management and Budget
OPIC	Overseas Private Investment Corporation
OTA	Office of Technology Assessment
PAO	Public Affairs Office
R & D	Research and Development
RDA	Research, Development, and Acquisition
RDT & E	Research, Development, Test, and Evaluation
RPV	Remotely Piloted Vehicle
RSA	Republic of South Africa
RUSI	Royal United Service Institution for Defence Studies (London)
SIPRI	Stockholm International Peace Research Institute
UK	United Kingdom
USA	United States Army
USAF	United States Air Force
USAINSCOM	United States Army Intelligence and Security Command
USDRE	Under Secretary of Defense for Research and Engineering
USI	United Service Institute (India)

USMC	United States Marine Corps
USN	United States Navy
USNI	U.S. Naval Institute
WHS	Washington Headquarters Services

In citations of Congressional hearings, 95-1 signifies the 95th Congress, First Session.

I.
INTRODUCTION

This research guide attempts to bring together references to the basic sources on worldwide military and strategic affairs. It is designed for students, researchers, journalists, peace activists, government workers, and others with a professional or scholarly interest in this field. The need for a research guide of this sort is dictated not only by the great importance and complexity of these issues, but also by the large number of sources. In this guide alone, over one thousand reference sources and six hundred periodicals are discussed.

USER'S GUIDE

Every effort has been taken to make this guide as useful to as many researchers as possible. Both basic and specialized sources are provided in each topic area, along with suggestions for advanced research. Most of the basic sources can be found in a large municipal or university library, while other sources may require a visit to specialized government libraries (many of which are identified in this guide). Throughout the guide, primary reliance is on recurring publications such as annual reports, periodicals, and serials. Often the most useful information appears in government documents and the specialized military press, documents that conventional research methods fail to address. By acquiring a familiarity with the major periodicals, Congressional hearings, and other key sources, one can readily become an expert on almost any issue covered here.

The first part of this guide, GENERAL INFORMATION SOURCES, describes the tools which are the most useful, most easily accessible, and broadest in scope. In this and the **Armed Forces and Weapons** section of WORLDWIDE MILITARY AND STRATEGIC AFFAIRS (Part V), most of the well-known sources are described. A separate section on U.S. GOVERNMENT DOCUMENTS AND INFORMATION describes the governmental documents system and introduces the hundreds of useful publications issued by government agencies. Most of the basic information on military and strategic affairs comes from government publications, many of which are fully described in this guide. Such publications obviously do not provide the interpretive information that may be desired by all researchers, but a familiarity with their content and use is absolutely essential. As a counterpart to such government publications, an effort has been made to include as many alternative sources as possible. Secrecy involving military programs and plans does not, in most cases, preclude the ability to conduct adequate research, especially on policy issues.

In doing research, one should remember some basic steps for both thoroughness and effectiveness. In each section, bibliographic sources are listed first. These are the keys to the sources not listed in this guide. Government and annual reference publications which provide the basic statistical data in each field are listed next. Considerable attention is also paid to periodicals which contain the most up-to-date information, and often the best insight and analysis. (Appendix A lists the periodicals referenced in the guide and indicates where some of these periodicals are indexed.) Finally, in most sections, some reference works and notable sources are discussed.

The layers of resources—the bibliographies and bibliographic material, government documents and reference works, periodicals and specific books—are presented with the reader in mind. This guide is by no means complete, and while there may be some disagreement about the value of one work compared to another, an effort has been made to cite as many recurring works as possible.

Finally, this research guide is not exhaustive. It has limitations for doing research in a number of areas which should be noted: military history, NASA and space, veterans, civil defense, Coast Guard, internal security, intelligence, and terrorism.

ACKNOWLEDGEMENTS

There are numerous government officials who aided me in my research for this book who I would like to thank. Gene Kubal, Al Hardin and Nancy Harvey of the Army Library were especially helpful, as were many of their fellow librarians. Phil Farris and Bettie Sprigg of the Department of Defense Public Affairs Office were also valuable sources. Scores of other public affairs officers in the Departments of Defense and State and the Arms Control and Disarmament Agency were also important sources, but are too numerous to thank personally.

I would especially like to thank Michael Klare and Delia Miller of the Institute for Policy Studies and David Johnson of the Center for Defense Information for their encouragement and assistance. And really most importantly, I would like to thank Alicia.

II. GENERAL INFORMATION SOURCES

There are numerous reference works that because of their worldwide coverage and authoritativeness provide the starting point for research in the general field of international relations and the specialized field of military and strategic affairs. The sources listed in this part should enable a researcher to perform secondary research to profile a country, to develop comparative data or to become familiar with any issue in military and strategic affairs. Together with the more specific resources covered in the later sections of this guide, one should be able to develop an expertise in the field. Many of the resources listed here are also the best specialized sources in certain areas. They will be referred to throughout the guide.

A. REFERENCE TOOLS

These basic reference tools are the starting point for any sort of research. Some are the basic tools of a political scientist and are available in any good university or public library. Some are specialized tools that military analysts consult and are rare in general library collections. One doing book or dissertation research should become familiar with them.

1. Guides to Research: A number of other research guides in this area provide additional resources not mentioned here. *The Information Sources of Political Science* (Frederick L. Holler, Santa Barbara, CA: ABC-Clio Press, 1975) is a five-volume comprehensive review of general sources and retrospective bibliographies in all areas of political science, law, international relations and international law, many not covered in this guide.

Two general international relations research guides are also recommended: *Guide to the Study of International Relations* (J.K. Zawodny, San Francisco: Chandler, 1966) and *The Study of International Politics: A Guide to the Sources for the Student, Teacher, and Researcher* (Dorothy F. Labarr and J. David Singer, Santa Barbara, CA: ABC-Clio Press, 1976). They cover many data sources as well as classic works in the field that could be consulted for background information on theories and history.

Two specialized military research guides are also worth mentioning. *National Security Affairs: A Guide to Information Sources* (Arthur D. Larson, Detroit, MI: Gale Research, 1973) contains an unannotated bibliography (with a key word index) of almost 4,000 titles covering the international system, U.S. national security, war, defense policy, organization and management theory and foreign country national security affairs. It also lists numerous other bibliographies, reference works, and periodicals, and contains an excellent list of U.S. and foreign research and educational organizations. *A Short Research Guide on Arms and Armed Forces* (Ulrich Albrecht, et al., London: Croom Helm, 1978) is not very useful as a bibliographic source but contains an excellent critical evaluation of some of the basic reference materials (incuding *The Military Balance* and the *SIPRI Yearbook,* discussed later).

Two somewhat old but specialized "power structure research" guides, *Power Research Guide* (AGITPROP/Europe-Africa Research Project, London: AGITPROP, 1970) and *NACLA Research Methodology Guide* (New York: NACLA, 1970) contain excellent and still relevant resources and tips for doing research into institutions and personalities, particularly members of ruling elites. They include methods for identifying and characterizing biographical, military, corporate and international data.

Other relevant, specialized research guides are described in other sections of this guide. There are research guides covering a number of other areas: U.S. government information and documents (Section IIIA1), business information (Section IVD1), and United Nations and international organizations (Section VF). Many regional research guides are also important. The Smithsonian Institution's new *Scholar's Guide* series is a directory of organizations, collections and publications in the Washington, D.C. area. Many volumes have been completed, including: *Russian/Soviet Studies, Latin American and Caribbean Studies, African Studies,* and *East Asian Studies.* Other regional guides are discussed in Section V:

International and Regional Politics in the Middle East and North Africa
The Modern Middle East: A Guide to Research Tools in the Social Sciences
Guide to Research and Reference Works on Sub-Saharan Africa
Sub-Saharan Africa: A Guide to Information Sources

Latin American Research in the United States and Canada: A Guide and Directory

The European Communities: A Guide to Information Sources

Sources of Information on the European Communities

The International Relations of Eastern Europe: A Guide to Information Sources

2. Bibliographies: Bibliographies and bibliographic resources are tremendous timesavers for conducting research. In a field like military and strategic affairs where a great deal of the important materials appear in periodicals, bibliographies and the abstract and indexing services covered in the next section are essential tools. All of the research guides mentioned in the preceding section are valuable bibliographic resources, particularly *National Security Affairs*. There are a number of services which should be used to find bibliographies. The *Bibliographic Index: A Cumulative Bibliography of Bibliographies* (New York: H.W. Wilson Co., 1937-1942-) is a triennial index of bibliographies published separately or contained in other works. It is available in almost every library and is the basic resource. A number of specialized military bibliographic resources also exist. The *War/Peace Bibliography Series* published by ABC-Clio Press has over 20 titles published through 1979, most of them referenced in this guide. The larger military libraries of the Department of Defense are also prolific bibliography producers. These bibliographies are normally limited to books in the library collection. All of the military libraries participate in the interlibrary loan system and qualified researchers are often allowed access to the collections with written permission. The five biggest libraries are:

Air Force Academy Library, Colorado Springs, CO 80840
Air University Library (Fairchild Library), Maxwell AFB, AL 36112
Army Library, Room 1A518, The Pentagon, Washington, D.C. 20310
U.S. Military Academy Library, West Point, NY 10996
U.S. Naval Academy Library, Annapolis, MD 21402

The Air University Library in particular produces numerous bibliographies, and the Army Library until recently produced a number of bibliographies that were sold through the Government Printing Office (GPO). The Imperial War Museum (Lambert Road, London, England) prepares a number of selected subject bibliographies which are available from the Printed Books Section.

Periodicals can be tremendous bibliographic resources, particularly for new books. *Aerospace Historian* and *Military Affairs* are two excellent resources for new books, studies, documents, etc., with extensive notes. A number of other magazines are also recommended for their book reviews and listings:

ADIU Report
American Political Science Review
Arms Control Today
Disarmament
Foreign Affairs
International Affairs (U.K.)
Journal of Politics
Orbis
Survival

Dissertations are another type of publication that the researcher might find useful. Dissertation indexes are available in almost every university library but there are also two specialized military listings. *Doctoral Dissertations in Military Affairs* (Allan R. Millett and B. Franklin Cooling, Manhattan, KS: Kansas State University Library, 1972) is the only comprehensive index. It is supplemented by a listing appearing in *Military Affairs* in April 1973, and in February every year thereafter. A bibliography of military dissertations is also published by University Microfilms, *The Military: A Catalog of Dissertations* (Ann Arbor, MI: Xerox, University Microfilms, 1976), which includes U.S. Ph.D. dissertations available for purchase from University Microfilms.

A number of international relations bibliographies are themselves comprehensive and can serve as resources. The Universal Reference System (URS) bibliographies require some time to learn how to use but are quite comprehensive: *URS/Political Science, Government and Public Policy Series* (Princeton, NJ: Princeton Research Publishing Co., 1965-69-) and *URS/International Affairs* (Princeton, NJ: Princeton Research Publishing Co., 1969-). They are multi-volume base sets with annual supplements. *International Relations Theory: A Critical Bibliography* (A.J.R. Croom and C.R. Mitchell, eds., New York: Nichols, 1978) contains a series of review articles covering the current and classic works of literature on such areas as theories of power, conflict and war, strategy and foreign policy analysis. *The Study of International Relations: A Guide to Information*

Sources (Robert L. Pfaltzgraff, Jr., Detroit, MI: Gale Research, 1977) is an annotated bibliography of classic books in foreign policy, military strategy, deterrence and the use of power. *The Foreign Affairs Bibliography* (Henry L. Roberts, et al., eds., New York: Bowker, 1973) is the latest of five ten-year bibliographies (this one 1962-1972) published by the Council on Foreign Relations. It is an annotated bibliography of over 9,000 books and documents.

Numerous general military and strategic affairs bibliographies have also been prepared. Two are particularly worth mention, *Quarterly Strategic Bibliography* (Wash, D.C.: American Security Council Education Foundation, 1977-) and *Strategic Studies Reading Guide* (1975 rev. ed.) (Laurence Motiuk, Ottawa: National Defense Headquarters, 1976). The *Quarterly Strategic Bibliography* was started in 1976 as *Current Bibliographic Survey of National Defense* (1976-77) and is a quarterly index of articles from about 200 periodicals, Congressional documents, and books and documents. It has an author and subject index but because of the presentation of material is more suited as a bibliographic resource than as an index. The *Strategic Studies Reading Guide* is a bibliographic project begun in 1970, with five annual supplements accumulated into a revised edition in 1975. It is an annually supplemented unannotated bibliography of materials related to strategic affairs.

Other bibliographies dealing with military and strategic affairs include:

National Security, Military Power and the Role of Force in International Relations: A Bibliographic Survey of Literature (DA Pam 550-19) (The Army Library, Washington, D.C.: GPO, Sept. 1976), 850 annotations and abstracts from books, periodicals and documents covering the period 1946-1976.

Crisis Forecasting and Crisis: A Critical Examination of the Literature (Richard W. Parker, McLean, VA: Decisions and Designs, Inc., Dec. 1976), prepared under contract for Defense Advanced Research Projects Agency (DARPA), bibliographic review articles and an unannotated bibliography in such areas as crisis forecasting and management, decisionmaking and indicator systems.

Economics and Foreign Policy: A Guide to Information Sources (Mark R. Amstutz, Detroit, MI: Gale Research, 1977), an annotated bibliography covering relations, politics and theories of international economic trade, aid, investments and multinational corporations including imperialism and the economics of war and defense.

American Defense Policy Since 1945: A Preliminary Bibliography (John Greenwood, Lawrence, KS: Univ. Press of Kansas, 1973), an unannotated retrospective bibliography on various aspects of U.S. defense policy.

Naval and Maritime History: An Annotated Bibliography (Robert G. Albion, Mystic, CT: Museum of American Maritime History, 1972), a complete general naval bibliography covering all periods in books and dissertations, supplemented annually by *Bibliography of Maritime and Naval History Articles* (Charles R. Schultz and Pamela A. McNulty, comps., College Station, TX: Texas A&M Center for Marine Resources, 1971-).

Air Superiority: Selected References (Air University Library, Maxwell AFB, AL, Aug. 1978), an 83-page bibliography on the role of air superiority and the lessons learned in Southeast Asia and the Middle East wars, and the use of air forces in the European theater.

An Aerospace Bibliography (Samuel Duncan Miller, comp., Office of Air Force History, Wash. D.C.: GPO, 1978), a selection of books and articles dealing with Air Force and aerospace technology, aerospace doctrine, and history, including a bibliography of bibliographies and a guide to reference works and documents collections.

Air Power and Warfare (USAFA Library (Special Bibliography Series #59) Colorado Springs, Sept. 1978), an unannotated bibliography on all aspects of air power and air warfare, including tactics and operations.

Military Institutions and the Sociology of War: A Review of the Literature with Annotated Bibliography (Kurt Lang, comp., Beverly Hills, CA: Sage, 1972).

Civil-Military Relations and Militarism: A Classified Bibliography Covering the United States and Other Nations of the World with Introductory Notes (Arthur D. Larson, comp., Manhattan, KS: Kansas State Univ. Lib., 1971).

Conscription: A Select and Annotated Bibliography (Martin Anderson, ed., Stanford, CA: Hoover Institution Press, 1976), almost 1,400 entries on military manpower procurement worldwide.

The Military in Developing Countries: A General Bibliography (Charles Kuhlman, comp., Bloomington, IN: Indiana University, Jan. 1971), 1,200 titles on the role of the military in developing countries.

3. Abstracts and Indexes: Magazines and journals are one of the best sources of military news and background material necessary to do original research. Magazines also provide the most up-to-date analysis of strategic and military issues. The list of periodicals dealing with military and strategic affairs in Appendix A is testimony to the wide range of general and specialized periodicals. The tools listed in this section are the keys to finding periodical material, whether it be from the mass media or the specialized military press.

Newspapers provide much raw data on events and often have important and original research or expose articles that contribute a great deal to the understanding of a particular subject. Newspaper indexes are themselves valuable research aids as they are sometimes so well annotated that referral to the newspaper is unnecessary. Newspaper indexes can also provide the data for chronologies and can afford a fast reference system for tracking news. The best newspapers for military and strategic affairs are the *New York Times, Washington Post* and *Christian Science Monitor.* The *Christian Science Monitor* often has more in-depth articles on defense than some of the other dailies. The *Washington Post* covers Washington developments well. The *Wall Street Journal* contains much information on defense contracts and industry and the economic aspects of defense. Most of the indexes listed below are available in large university and public libraries.

New York Times (NYT Index, New York, 1851-1906/1912-), biweekly with annual cumulations.

Washington Post (Washington Post Index, Newspaper Indexing Center, Wooster, OH: Bell & Howell, 1972-). The *Washington Post* is also indexed in the *Federal Index* (see Section IIIA2).

Los Angeles Times (The Los Angeles Times Index, Newspaper Indexing Center, Wooster, OH: Bell & Howell, 1972-).

Chicago Tribune (The Chicago Tribune Index, Newspaper Indexing Center, Wooster, OH: Bell & Howell, 1972-).

Wall Street Journal (The Wall Street Journal Index, New York: Dow Jones, 1958-1967-), monthly with annual cumulations.

Christian Science Monitor (Index to the Christian Science Monitor, Boston: CSM, 1960-), monthly with semiannual and annual cumulations.

The Times (London) *(Index to the Times,* John Gurnett, ed., Reading, England: Newspaper Archive Developments, 1906-).

Magazines and journals can also be a source of information, although the data provided is rarely sufficient to contribute to original research. A number of these magazines report defense issues regularly. *The Reader's Guide to Periodical Literature*

(New York: H.W. Wilson Co., 1900-1905-) indexes popular, nontechnical periodicals. *The Vertical Index: A Subject and Title Index to Selected Pamphlet Materials* (New York: H.W. Wilson Co., 1932-) may also prove useful listing new pamphlets, monographs, leaflets and similar material.

A number of abstracts and indexes exist covering the broad areas of political science and public administration. These cover the specialized journals in these fields, may of which are not referenced in this guide but which contain occasional articles about defense issues and military and strategic affairs.

Public Affairs Information Service Bulletin (PAIS) (Robert S. Wilson, ed., New York: PAIS, 1914-1951-), biweekly with quarterly and annual cumulatives covering popular works and general political science, public administration and general public policy, with books, monographs, government documents and periodical articles indexed.

Sage Public Administration Abstracts: An International Information Service (Beverly Hills, CA: SAGE, 1973-), quarterly abstracts of public administration and management journals covering policy making, administration, bureaucracy, and planning.

Social Science Index (New York: H.W. Wilson Co., 1966-), monthly with quarterly cumulatives of journals in the social sciences covering theoretical aspects of the social sciences.

ABC Poli Sci: A Bibliography of Contents: Political Science and Government (Santa Barbara, CA: ABC-Clio Press, 1969-), five times a year with annual index; indexes the article titles and the table of contents of 300 political science and government journals, including a subject index in such areas as arms control and disarmament, defense, militarism, intervention, war, and weapons.

Two international tools cover the middle area between political science and the specialized military science publications. *International Political Science Abstracts* (Paris: International Political Science Association, 1951-) is a bimonthly service of over 150 English and foreign language publications, with a special section on "international relations." The other index is *Public International Law: A Current Bibliography of Articles* (Max Planck Institute for Comparative Public and International Law, New York: Springer-Verlag, 1974-), a semiannual unannotated international bibliography and index of works in international law including war and armed conflict, international organization, and sea, air and space law. This service should be available in a good law library.

Abstracts and indexes of military periodicals are no fewer in number than those in other fields but are much rarer in general library collections. The most common index is the *Air University Library Index to Military Periodicals* (Maxwell AFB, AL: Air University Library, Oct./Dec. 1949-), a quarterly (with annual cumulations) subject index of 67 English language military periodicals. Most of these periodicals are quite common and all of them are referenced in Appendix A. The *Air University Index* is by no means complete, however, and there are a number of major military magazines which it does not cover. Another common index is *Peace Research Abstracts Journal* (PRAJ) (Dundas, Ontario, Canada: Peace Research Institute, 1964-), a monthly, well-annotated abstracts service that is coded in a unique subject categorization system, but one that is parallel to most research topics. The ten areas are:

  I. The Military Situation
 II. Limitations of Arms
III. Tension and Conflict
 IV. Ideology and Issues
  V. International Institutions and Regional Alliances
 VI. Nations and National Policies
VII. Pairs of Countries and Crisis Areas
VIII. International Law, Economics and Diplomacy
 IX. Decision Making and Communications
  X. Methods and Miscellaneous

PRAJ has an annual subject index and each issue has an author index. A new abstracts service produced by the Center for Naval Analysis (CNA) is a must for any research in the naval and maritime fields. *Naval Abstracts* (Alexandria, VA: CNA, 1978-) is a quarterly (with cumulative index) service of short abstracts of articles from 273 journals (English and foreign language) dealing with worldwide naval matters, including broad strategic issues impacting on the navies of the world and naval sciences. With a subject and author index, *Naval Abstracts* proves to be a far better (although more limited in scope) tool than the *Air University Index*. All of the English language magazines abstracted in *Naval Abstracts* are listed in Appendix A. *Abstracts of Military Bibliography* (Buenos Aires, Argentina, 1967-), although very selective, is a service which the researcher should find useful since it abstracts some periodicals not covered by the other military abstracts and indexes listed here. It is a quarterly service abstracting about 60 articles, books and monographs per issue and indexing about 120 other titles in such areas as international and internal armed conflict, modern military organizations and institutions, weapons, security systems and policies, and arms control and disarmament. It selects the major articles from about 125 military magazines (including many Latin American military magazines) and reproduces the Table of Contents of about ten periodicals. Another index, also selective in nature, is produced in England, *Military Science Index: A Monthly List of Papers of Military Interest from Current Periodicals* (Schrivenham, England: Royal Military College of Science Library, 1962-), with annual cumulations. It indexes articles from a wide variety of sources which the library selects on the basis of the needs of the Royal Military College.

Finally, there are a number of other abstracts and indexes which the military researcher should become familiar with. There are abstracts and indexes covering U.S. government publications which are essential for research (see Section IIIA2), government technical information and reports (see Section IIIC4), business information and business and trade periodicals (see Section IVD1), and United Nations documents and publications (see Section VF). There are also two regional abstracts that are important tools: *The Middle East: Abstracts and Index* (see Section VB5) and *African Abstracts* (see Section VB1).

4. Dictionaries: The terminology and jargon of military and strategic affairs often intimidates the researcher. Like any other specialized field, the military has developed a language which meets the needs of clear exposition, organization and policy. And like any bureaucracy, the military has adopted the use of acronyms and abbreviations to such an extent that military writings and official documents are sometimes incomprehensible. Therefore the researcher should become familiar with a number of specialized dictionaries. A comprehensive historical bibliography of dictionaries, *A Bibliography of Encyclopedias and Dictionaries Dealing with Military, Naval and Maritime Affairs, 1577-1971* (4th Ed.) (Craig Hardin, Jr., comp., Houston, TX: Fondren Library, Rice University, 1971) contains reference to many dictionaries not listed below (including many dictionaries of military terminology in foreign languages). The dictionaries listed here are of two types, commercially produced and government produced. They cover conceptual terminology, definitions, official word usage and acronyms.

The most useful dictionary of conceptual terminology is *A Dictionary of Modern War* (Edward Luttwak, New York: Harper & Row, 1971), which has over 500 entries covering both technical and conceptual terms, weapons systems, operational terms and strategic jargon. A newer dictionary, *Words and Arms: A Dictionary of Security and Defense Terms (with supplementary data)* (Wolfram F. Hanreider and Larry V. Buel, Boulder, CO: Westview, 1979), covers terms and concepts and draws heavily from Congressional Budget Office publications for definitions and data. *The Encyclopedia of Modern War* (Roger Parkinson, New York: Stein & Day, 1977) covers the concepts of warfare and is more historical than the two mentioned above. Finally there are two government produced dictionaries of contemporary terms which contain the definitions of commonly used jargon (i.e., Nuclear Club, trilateralism, basket three, and slang). *The International Relations Dictionary* (Dept. of State Library, Wash., D.C.: GPO, 1978) and *SALT Lexicon* (Arms Control and Disarmament Agency, Wash., D.C.: GPO, 1974) are both available for a nominal fee from the Government Printing Office (GPO).

The two naval terms dictionaries, *Naval Terms Dictionary* (4th Ed.) (John V. Noel and Edward L. Beach, Annapolis, MD: USNI, 1978) and *Jane's Dictionary of Naval Terms* (Joseph Palmer, London: Macdonald and Jane's/New York: Hippocrene Books; 1975/1976) are similar, with the Jane's dictionary having a predominance of British naval terms. *Jane's Dictionary of Military Terms* (P.H.C. Hayward, London: Macdonald and Jane's, 1975) also has a British predominance and contains many archaic terms. *The Dictionary of Weapons and Military Terms* (John Quick, New York: McGraw Hill, 1973) is a better general source with many technical terms, code names and slang.

Two specialized dictionaries of Soviet military terminology and comparative U.S.-Soviet usage are highly recommended. The *Comparative Dictionary of U.S.-Soviet Terms* (Wash., D.C.: DIA, DDI-2200-33-77, 1977) and *Dictionary of Basic Military Terms* (USAF, Soviet Military Thought Series #9, Wash., D.C. (Moscow, 1965): GPO, 1976) are two unclassified intelligence publications which can prove indispensable.

The official dictionaries of the Department of Defense define many operational and logistic terms and are reference sources used by every military officer. The *Department of Defense Dictionary of Military and Associated Terms* ("JCS Pub 1") (DOD, Wash., D.C.: GPO, 1979) is the basic source and is often quoted in scholarly articles and the media. Other less known dictionaries of service-peculiar terminology are *The United States Air Force Dictionary* (Woodford A. Heflin, ed., Wash., D.C.: GPO, 1956), *U.S. Air Force Glossary of Standardized Terms* (AFM 11-1, Vol. I) (USAF, Wash., D.C.), *Dictionary of*

U.S. Army Terms (AR 310-25) (USA, Wash., D.C.), and *Naval Terminology* (NWP-3) (USN, Wash., D.C.). The *NATO Glossary of Military Terms and Definitions for Military Use* (in French and English) (AAP-6) (NATO, Military Agency for Standardization, 1977) utilizes terms from JCS Pub 1.

Dictionaries of acronyms and abbreviations can prove to be quite troublesome. Since there is not a comprehensive dictionary of defense acronyms, sometimes a number of dictionaries need to be consulted. The commercial general abbreviations dictionaries, *Acronyms, Initialisms and Abbreviations Dictionary* (6th Ed.) (Ellen T. Crowley, ed., Detroit, MI: Gale Research, 1978) and *Code Names Dictionary* (Frederick G. Ruffner, ed., Detroit: Gale Research Co., 1963) are not very useful for military acronyms. The *Dictionary of Naval Abbreviations* (2nd Ed.) (Bill Wedertz, Annapolis, MD: USNI, 1977) is very good for U.S. Navy acronyms and the *Code Name Handbook* (Greenwich, CT: DMS, Inc.) (annual) is excellent for equipment and systems abbreviations and code names. The Army and the Air Force also have official acronyms dictionaries, *Authorized Abbreviations and Brevity Codes* (AR 310-50) (USA, Wash., D.C.) and *Air Force Manual of Abbreviations* (AFM 11-2) (USAF, Wash., D.C.) which are continually updated and are comprehensive for service-wide acronyms.

B. DATA SOURCES

This section presents the basic reference sources for finding statistical data on countries of the world, armed forces, military expenditures and the arms trade. It should be used in conjunction with other references and sections in this guide. It is a starting point for comparative data, and for the most authoritative facts on quantifying military and strategic affairs. These sources are the most widely quoted and thus should be familiar to any researcher.

1. General Data: Data on countries of the world are found in a number of annual almanacs and factbooks. This data includes mostly economic and demographic statistics needed for strategic research and brief profiles of a country's history, government, politics, people and relations. The resources listed here and many others are listed in an excellent reference book, *Encyclopedia of Geographic Information Sources* (3rd Ed.) (Paul Wasserman, ed., Detroit, MI: Gale Research Co., 1978), which is a geographic breakdown (by region and country) of reference sources (mostly geared towards business research) including yearbooks, bibliographies, biographical information, directories, periodicals, abstracts and indexes and statistics. Another reference source for finding specific country data is *Government Organization Manuals: A Bibliography* (Vladimir M. Palic, Library of Congress (LC), Wash., D.C.: 1975) which is a useful listing of manuals and reference works on individual countries of the world. The U.S. Army *Area Handbooks* (DA Pam 550-XX series) (USA, Wash., D.C.: GPO) (produced under contract by Foreign Area Studies, American University), and the State Department *Background Notes on the Countries of the World* (Dept. of State, Wash., D.C.: GPO) are also excellent basic reference sources. The Army *Area Handbooks* (now being produced as *Country Studies*) are the best sources for some countries of the world, including most of the Third World, and excellent background sources. They are available from the GPO and are updated periodically. The general sources for more capsulized information are:

The International Yearbook and Statesmen's Who's Who 19__ (East Grinstead, England: Kelly's Directories Ltd., 1864-1919-), country profiles, international and regional organizations and Who's Who.

Political Handbook of the World 19___ (Arthur S. Banks, ed., New York: McGraw Hill, 1927-), a review of world and regional issues, country profiles (including foreign relations), and international organizations.

National Basic Intelligence Factbook (CIA, Wash., D.C.: GPO), semiannual unclassified version of SECRET Factbook prepared by the CIA of basic statistics on the people, government, economy, communications and military forces, printed commercially by Gale Research Co. of Detroit as *Handbook of the Nations*.

The Europa Yearbook 19__: A World Survey (London: Europa Publications, 1926-1959-), a two-volume comprehensive review of international organizations, and country profiles with much background information and statistical material.

The Statesman's Yearbook 19__-__ (New York: St. Martin's Press), international organizations and country profiles including information on defense in each country (something not covered by most yearbooks).

Handbook of Economic Statistics 19__ (CIA, Wash., D.C.: National Technical Information Service (NTIS)), a compilation and presentation of statistical data on various aspects of international economic trade, aid, energy, and natural resources, covering selected western and communist countries.

2. Armed Forces: General data on armed forces of the world is contained in a number of reference works. This data includes information on force structure, military personnel, and inventory of equipment. Some of the sources listed below are considered the bibles of military information but as this guide indicates, they are by no means the totality of information available. Some of the sources have serious limitations, some of which are discussed in *A Short Research Guide on Arms and Armed Forces* mentioned earlier.

The most reputable source, *The Military Balance 19__-__* (London: International Institute for Strategic Studies, 1959-60-), contains no explanatory text for its country profiles and lists no sources. It is, however, the best annual compilation of material on military forces and regional power. It is supplemented by special comparative tables and essays. *The Military Balance* is also reprinted in the December issue of *Air Force Magazine*. It is the most widely quoted source of information on military organizations and formations. Another highly regarded data source is the *SIPRI Yearbook (World Armaments and Disarmament)*, which presents much data on the state of the arms race and worldwide military expenditures, but none on armed forces. *The Defense and Foreign Affairs Handbook 19__* (George R. Copley, ed., New York: Franklin Watts, 1978-) (annual) and *The Almanac of World Military Power* (4th Ed.) (Trevor N. Dupuy, ed., et al., New York/London: R.R. Bowker Co., 1979) (biennial) present country-by-country military information. They both cover military forces, with the *Almanac* discussing regional and country strategic situations and military assistance and the *Handbook* discussing defense production and industry as well as the media. The *Handbook* also contains an appendix with an extensive list of defense manufacturers, producers and suppliers and a Who's Who. Both of these sources provide amplifying data to *The Military Balance*. Another book found in many libraries is the *Reference Handbook of the Armed Forces of the World* (4th Ed.) (Robert C. Sellers & Associates, New York: Praeger, 1966-), which presents statistical data on country military forces.

An information service compiled by Defense Marketing Service, Inc. (DMS), *DMS Market Intelligence Report: Foreign Military Markets, Vol. I: NATO/Europe, Vol. II: Middle East/Africa, Vol. III: South America/Australasia* (Greenwich, CT: DMS, Inc.), is an excellent source on 86 countries of the world. The information is compiled from the perspective of providing defense industry and governments information on the arms needs of a particular country. These volumes are comprehensive examinations of military organization, posture, budget, arms purchases, force structure, manufacturing capability, procurement organization and policy, future needs and current suppliers.

These are the basic sources of information on the armed forces of the world. Many other sources of a more specialized nature are listed in Section VA, *Armed Forces and Weapons.*

3. *Military Expenditures:* Military expenditures are often used as a yardstick to compare countries and assess a country's commitment to defense. Although this is by no means a complete picture of a country's military might, the researcher must become familiar with the basic sources which compile data on military expenditures and the allocation of resources to the military. Most of the sources listed in this section also break down and compare military and military-related personnel and economic indicators bearing on the overall investment and influence of the military in the local economy. The basic sources on military expenditures are two annual publications, *World Military Expenditures and Arms Trade, 19__-19__* (ACDA, Wash., D.C.: GPO, 1966-) and *World Military and Social Expenditures 19__* (Ruth Leger Sivard, Leesburg, VA: World Priorities Publications, 1974-). They present worldwide statistical data on military expenditures, armed forces and the use of resources for military and social purposes. Other authoritative sources for data include many of the sources cited above under **Armed Forces** including *Military Balance* and the *SIPRI Yearbook.* The *SIPRI Yearbook* contains much data on the state of the arms race, military expenditures and the arms trade that is the result of independent methodological research. There are numerous sources for quantitative data on the arms race, as well as many sources which provide background on the subject. Three bibliographic and research sources are *Review of Research Trends and an Annotated Bibliography: Social and Economic Consequences of the Arms Race and of Disarmament* (International Peace Research Association, New York: UNESCO, 1978), "The Use of Resources for Military Purposes: A Bibliographic Starting Point," by Manne Wangborg in *Bulletin of Peace Proposals,* March, 1979, and the later expanded *Disarmament and Development: A Guide to Literature Relevant to the United Nations Study* (Manne Wangborg, Stockholm: Forsvarets Forskningsanstalt, 1980?). Another bibliographic source already mentioned is *Economics and Foreign Policy: A Guide to Information Sources.*

Two valuable studies which also include extensive bibliographic material are *Resources Devoted to Military Research and Development* (SIPRI, Stockholm: Almqvist & Wiksell, 1972), which discusses the resources devoted to military research and development in twenty-four industrialized countries, and *The Economics of Third World Military Expenditure* (David K. Whynes, Austin: Univ. of Texas Press, 1979), a study of military institutions and expenditures and their effects in developing countries.

Finally, the United Nations has done a number of studies in this area as a result of resolutions and proposals to limit military spending. They discuss the dynamics of the arms race and attempt to define the scope of military spending to develop a standardized international reporting system. Three of the studies are:

Economic and Social Consequences of the Armaments Race and Its Extremely Harmful Effects on World Peace and Security: A Report of the Secretary-General (U.N. General Assembly (A/32/88), New York: U.N., Aug. 12, 1977).

Economic and Social Consequences of the Arms Race and of Military Expenditures (U.N., Centre for Disarmament, New York: U.N., 1978).

Reduction of Military Budgets: Measurement and International Reporting of Military Expenditures (U.N., Dept. of Political and Security Council Affairs (E.77.1.6.), New York: U.N., 1977).

4. The Arms Trade: The worldwide trade in arms and its effects on international security and regional balances of power is another area which receives much research attention. Much statistical data is available on the arms trade and much has been written on the subject. The sources for quantifying the extent of this trade are covered in this section. U.S. military assistance is covered comprehensively in Section IVG4. Many of the same sources listed above under **Armed Forces** and **Military Expenditures** apply to the arms trade. Many contain gross statistics on arms transactions. The *SIPRI Yearbook* is the most authoritative source of information on the arms trade with its "arms trade registers" which review the major transactions of each year. A compilation of these registers has also been published as *Arms Trade Registers: The Arms Trade with the Third World* (SIPRI, Cambridge, MA: MIT, 1975) which includes listings of transactions to January 1974. *The Military Balance* also contains listings of transactions but they are not as comprehensive as the SIPRI listings. The *Almanac of World Military Power* describes military aid for each country as part of its profiles and *World Military Expenditures* and *World Military and Social Expenditures* provide much tabular data. Two CIA publications, *Communist Aid Activities in the Non-Communist Less Developed Countries 19__* (CIA, Wash., D.C.: NTIS, 1976?-) (annual), and *Arms Flows to LDCs: U.S.-Soviet Comparisons, 1974-1977* (CIA, (ER 78-10494U), Wash., D.C.: Library of Congress, Nov. 1978), are also excellent sources, with *Communist Aid Activities* providing an annual survey of economic and military aid programs and country surveys of recipients. A comprehensive listing of recent U.S. transactions is *Arms Trade Data,* compiled by the Institute for Policy Studies, and *Foreign Military Sales and Military Assistance Facts,* an annual report of the Defense Security Assistance Agency of the Department of Defense.

Current transactions are reported in many military periodicals. Arms transactions between the NATO countries receive much attention because of the volume of sales and their political significance. Large transactions of the United States are usually well covered in the media. The magazines listed covering worldwide military affairs (see Table 2) are the best sources for raw data on current transactions. Some report transactions as a regular feature. *Air International,* for instance, has a regular section, "Airscene," which is a monthly review of worldwide military contracts and transactions. Many of the newsletters, particularly DMS, Inc. services, are also excellent sources of raw data.

C. DESCRIPTIVE SOURCES

The reference works cited in the last section were primarily sources of quantitative data and background material. Some are essential reference books that will be referred to frequently. The data they provide helps clarify what is in the mass media and provides the resources necessary to develop a position or perform documented research. While some of these sources—particularly *The Military Balance* and the *SIPRI Yearbook*—are both data and descriptive sources, some are primarily descriptive, reporting current events and trends, providing background on personalities in the news, and documenting and reviewing world, military and strategic affairs. These, and the periodicals for following current developments in military and strategic affairs, are discussed in this section. Similar descriptive sources on regional and country issues are mentioned here and discussed in Section V.

1. Events: Three services compile information on worldwide current events. Similar in coverage, they report foreign affairs, politics and domestic issues. All three are available in most large libraries.

Facts on File: World News Digest with Cumulative Index (New York: Facts on File, Inc., 1940-), a weekly report with monthly, semiannual and annual indexes of world events. No sources are given, but previous reports on the same subject are referenced.

Keesing's Contemporary Archives: Weekly Diary of World Events with Index Continually Kept Up-to-date (Bristol, England: Keesing's, 1931-), a weekly, with cumulative biweekly, quarterly, annual and biennial index news report. *Keesing's* is partial to the British Commonwealth and its reporting is comprehensive in this area. Covering foreign and domestic policy, including sources and texts of significant government documents and speeches, this is an excellent research tool.

Deadline Data on World Affairs (Greenwich, CT: Deadline Data, 1956-), a weekly card service (issued on 5x8 cards) of news and background information categorized by country, with a monthly cumulative supplement called *On Record.*

These three services can be timesavers for compiling a chronology, researching or summarizing a current or historic event. A number of regional news services also serve as research aids. They are described in Section V. Newspaper indexes can also be used for this purpose, particularly the *New York Times Index,* which often provides enough information in its annotations to eliminate the need of going to the newspaper.

Prepared chronologies are also valuable aids to a researcher and a number are compiled in the military and strategic fields. The best all-around chronology is *Chronologies of Major Current Developments in Selected Areas of International Relations* (Prepared by the Congressional Research Service, Library of Congress, for the Committee on Foreign Affairs, U.S. House of Representatives, Wash., D.C.), a monthly selective annotated chronology of international events of interest to the Congress. Excellent military and strategic chronologies appear annually in *RUSI and Brassey's Defence Yearbook,* and *Strategic Survey,* and a chronology of U.S. foreign policy appears annually in *United States in World Affairs.* A chronology of conflicts and terrorism also appears annually in *Annual of Power and Conflict.*

A number of journals also contain excellent chronologies,

with events in foreign policy, international conflicts and military affairs included. The best chronology appearing in a military magazine is in the annual "Naval Review Issue" (May issue) of *USNI Proceedings*. This is a comprehensive review of naval and maritime events of the preceding year. Other magazines which include regular chronologies are:

Africa Report
Current History
Disarmament
Free China Review
International Affairs (Poland)
Japan Quarterly
Korea and World Affairs
The Middle East Journal
Pakistan Horizon

A number of annuals referenced later in the regional sections also contain chronologies. They include:

Africa Yearbook and Who's Who
Asia Yearbook
The Caribbean Yearbook of International Relations
China Facts and Figures Annual
U.S.S.R. Facts and Figures Annual

2. Personalities: Biographic information on personalities in the news as well as listings of world leaders can also be found in some basic reference sources. The *Biographic Index* (New York: H.W. Wilson, 1947-) is the basic source for articles written on personalities receiving public attention. Listings of world leaders and ministers of government are contained in two publications, *Almanac of Current World Leaders* (Los Angeles, CA: Llewellyn, 1958-), a monthly magazine which provides brief biographies and quarterly listings of ministers, leadership changes and events leading to changes in government, and *Chiefs of State and Cabinet Members of Foreign Governments* (CIA, National Foreign Assessment Center, Wash., D.C.: NTIS), a monthly listing. The various *Who's Whos* also provide listings and brief biographic profiles. An index of these is the *Encyclopedia of Geographic Information Sources,* which lists biographic information sources. Some of the annuals mentioned in this guide contain Who's Whos:

The International Yearbook and Statesmen's Who's Who South of the Sahara 19__-19__

Africa Yearbook and Who's Who
*The Middle East and North Africa 19_-19_*
The Pacific Islands Year Book
*Far East and Australasia 19_-19_*

In addition to the general biographic sources listed above, there are a few specialized military sources. *World Military Leaders* (Paul Martell and Grace P. Hayes, eds., New York: Bowker, 1974) (also published in England by Macdonald and Jane's as *World Defence Who's Who* is a unique resource. The *Defense and Foreign Affairs Handbook* contains a Who's Who which updates much of that data. All military magazines report appointments of general officers, promotions and command changes. The *Army Quarterly and Defence Journal* is an excellent source for information about British military officers.

Information on American generals is contained in a number of periodicals. The current Marine Corps and Navy general officers and admirals are listed in the "Naval Review" issue (May) of *USNI Proceedings*. *Army* magazine's annual "Green Book" (October) lists Army general officers. *Army Times, Navy Times,* and *Air Force Times* report promotions, command changes and retirements every week, and a number of other magazines (*Armed Forces Journal International, Army* and *Air Force Magazine*) also report general officer changes. Official biographies of all general officers and civilian officials are compiled by the public affairs offices of the Department of Defense and the military services, and are available upon request. In addition, the personnel offices of the services compile listings of general officers, also obtainable from the public affairs offices. The National Guard Bureau compiles biographies in a continually updated set, "General Officers of the Army and Air National Guard."

3. *Reviews:* Reviews of world, military and strategic affairs are excellent resources for research. A number of annuals analyze events and trends in the previous year in a number of different subject areas. These can be especially valuable for understanding complex issues. In addition to the annual reviews, the numerous military, strategic and international relations journals carry in-depth articles on current issues. Many of the periodicals listed in Appendix A are also listed in this section, in three catgories: general international relations, strategic affairs and military sciences.

There are many reviews of world affairs and international relations. Some of these listed below provide statistical and background data on the countries of the world and have been mentioned under Section IIB1, **General Data.**

The Year Book of World Affairs 19___ (London, Institute of World Affairs, Boulder, CO: Westview, 1946-), articles on various issues of general world affairs (regional issues, energy, superpower relations, geopolitics).
The Annual Register: World Events in 19___ (H.V. Hudson, ed., London: Longman/New York: St. Martin's Press, 1758-), a review of the year's activities in the U.K., the Commonwealth, international organizations and nations of the world.
Global Political Assessment (New York: Columbia University Research Institute on International Change, 1975-), a semiannual survey of world affairs including regional surveys and the East-West military balance.
Survey of International Affairs (London: Royal Institute of International Affairs, 1920-), essays summing up the previous year in terms of the major world political and military events.
The Yearbook of World Policy (New York: Praeger, 1957-), an annual review.
Documents on International Affairs (London: Royal Institute of International Affairs), an annual collection of significant documents, treaties, and agreements affecting international affairs.
Indian Yearbook of International Affairs (Madras, India: Univ. of Madras, 1952-), an annual review of international affairs and Indian foreign policy.
Japan Annual of International Affairs (Tokyo: Japanese Institute of International Affairs, 1964-), an annual review of international affairs and Japanese foreign policy.
The World This Year (New York: Simon & Schuster, 1971-), an annual review.

In addition to these general international relations reviews, there are two annual reviews of international Communist affairs, subversion and internal security issues:

Annual of Power and Conflict 19__-__: A Survey of Political Violence and International Influence (7th Ed.) (Brian Crozier, ed., London: Institute for the Study of Conflict, 1971-), a 90-country review of internal and external threats, terrorism and militancy, including a chronology of events.
Yearbook of International Communist Affairs (Milorad M. Drachovitch, Stanford, CA: Hoover Institution, 1967-), country reviews of international and political communist activities.

Finally, many international relations and world affairs journals report general developments in international relations and carry occasional articles on military and strategic affairs. They are listed in Table 1. A number of peace research journals also carry many articles on general international relations.

Of the reviews of strategic and military affairs, three are most prominent: *World Armaments and Disarmament: SIPRI Yearbook 19__*(SIPRI, London: Taylor & Francis Ltd., 1968/69), *RUSI and Brassey's Defence Yearbook 19__/__*(Royal United Service Inst., ed., New York: Crane, Russak, 1971-), and *Strategic Survey 19__* (London: International Institute for Strategic Studies, 1971-). The *SIPRI Yearbook,* prepared by the staff of the Stockholm International Peace Research Institute (SIPRI), is an independent and unbiased assessment of the state of the arms race, developments in arms control and disarmament, world military expenditures, and arms production and trade. Regular features include the "Arms Trade Registers," which list major arms transactions of the year. Each *SIPRI Yearbook* contains not only a wealth of statistical data, but a number of analytic articles of current interest on subjects such as military forces and technology, arms control, nuclear weapons, and nonproliferation. The *SIPRI Yearbook* is well footnoted and includes sections on research methodology.

RUSI and Brassey's Defence Yearbook is a three-part yearbook which includes a review of military and strategic affairs, a weapons and technology review, a bibliography and a chronology. About ten articles by various scholars assess different aspects of the strategic situation in Part I. Part II is an excellent analysis of the current and future state of weapons technology. It is also sold separately as *Weapons Developments.* The third part is a review of defense literature and a chronology.

The *Strategic Survey* was begun in 1971 as a counterpart to the International Institute for Strategic Studies' *The Military Balance*. It is a concise review of the strategic situation in all regions of the world, of the superpower relationship, and of arms control and new factors affecting world security (including the influence of weapons technology). The *Strategic Survey* contains a chronology which is the most concise and comprehensive review of strategic events and issues. The *Strategic Survey,* along with *The Military Balance,* is the best investment any researcher can make in this field.

In addition to the general reviews of world affairs and the military and strategic annuals listed in this section, a number of more specialized descriptive reviews are worth mentioning here (and are described in detail later in this guide):

TABLE 1:
GENERAL INTERNATIONAL RELATIONS JOURNALS

Alternatives: A Journal
  of World Policy
Aussen Politik (English edition)
British Journal of
  International Studies
Comparative Political Studies
Comparative Politics
Cooperation and Conflict
The Fletcher Forum
Foreign Affairs
Foreign Policy
India Quarterly
Intelligence Digest World Report
International Affairs (Poland)
International Affairs (U.S.S.R.)
International Affairs (U.K.)
International Interactions
International Journal
International Organization
International Problems
International Relations
International Studies
International Studies Quarterly
Jerusalem Journal of
  International Studies
Orbis
Policy Review
Problems of Communism
Review of International Affairs
The Round Table
Spettatore Internazionale
Studies in Comparative
  Communism
Survey
Swiss Review of World Affairs
World Affairs
World Politics
The World Today

Indian Armed Forces Yearbook
Middle East Contemporary Survey
Middle East Annual Review
African Contemporary Record
The Caribbean Yearbook of International Relations
Pacific Defense Reporter Yearbook
Asia Yearbook
Military Year Book
The Chanakya Defence Annual
Soviet Armed Forces Review Annual

The general international relations journals have strategic and military counterparts. Table 2 contains a list of the periodicals in these fields which regularly review military and strategic affairs in scholarly and analytic articles.

TABLE 2:
STRATEGIC AND MILITARY JOURNALS

Strategic:
ADIU Report
AEI Foreign Policy and
  Defense Review
Arms Control Today
The Bulletin of the
  Atomic Scientists
Canadian Defense Quarterly
Comparative Strategy
Conflict
Conflict Studies
International Security
International Security Review
Journal of Strategic Studies
Orbis
RUSI Journal
Strategic Digest
Strategic Review
Strategic Studies
Survival
The Washington Quarterly
World Politics

Military:
Air Force Magazine
Air University Review
Armed Forces and Society
Armed Forces Journal
  International
Army
Army Quarterly and
  Defence Journal
The Australian Journal
  of Defence Studies
Defence Force Journal
Defense Monitor
The Hawk
History, Numbers and War
Military Review
NATO's Fifteen Nations
Naval War College Review
Parameters
RUSI Journal
USI Journal
USNI Proceedings

III.
U.S. GOVERNMENT DOCUMENTS AND INFORMATION

As a group, U.S. government publications are the greatest source of information on military and strategic affairs. The reference materials prepared and disseminated by various branches of the government provide the Congress with the materials necessary to make intelligent decisions on defense and foreign policy. Much of this material is unclassified and in the public domain. The system, however, is complicated and parts of it poorly organized for making information available to the public at large. The Defense Department is one of the biggest culprits, often making it difficult (though not impossible) to get documentation and information. In this section, both the system of government publications and an overview of the publications available are discussed.

A. INTRODUCTION

The tools required to research government information and publications are the same as in the other fields described in this guide: guides to research, indexes and abstracts, and reference manuals. The guides to research are essential supplements to the brief introduction here. Many types of publications are prepared by many agencies. One researching a specific area or requesting information from a government agency will want to find more in-depth descriptions of the agency's operations and publishing system. The various types of publications, from Defense Department reports and internal information to Congressional hearings to CIA intelligence working aids, require some preliminary research to understand the agencies and the types of information they regularly publish or can make available to the public.

The first step in using government documents for reference is understanding the Government Printing Office (GPO) and the responsiblities of the Superintendent of Documents. These are described in the research guides in the next section. While the GPO prints and sells most government documents, a great deal of the material in the military and strategic fields is not printed or sold by the GPO. CIA publications, many Defense Department publications, and Congressional Research Service (CRS) and General Accounting Office (GAO) publications are only available from the originating agency or through special channels. Therefore, in this field one cannot rely on the GPO and its *Monthly Catalog of U.S. Government Publications.*

1. Guides to Research: The GPO and Superintendent of Documents system, government information policy, publications, legislation and information are discussed in a number of excellent reference books. *Introduction to U.S. Public Documents* (2nd Ed.) (Joe Morehead, Littleton, CO: Libraries Unlimited, 1978) is the best and most up-to-date discussion of these subjects. It is a detailed guide to government publications, guides, indexes, and agency publishers, covering the executive, legislative and judiciary branches. The classic work on the subject is older, though still valuable as a basic guide to government publications, *Government Publications and Their Use* (2nd rev. ed.) (Laurence F. Schmeckebier and Roy B. Eastin, Wash., D.C.: Brookings Institution, 1969). Other research guides worth looking at, specifically for their coverage of the military and strategic fields, are:

Government Reference Books_/_: A Biennial Guide to U.S. Government Publications (Allan Edward Schorr, comp., Littleton, CO: Libraries Unlimited, 1978), good coverage of military and foreign policy titles.

Guide to U.S. Government Publications (John L. Andriot, Arlington, VA: Documents Index, 1962-), a regularly updated guide to serials, periodicals and reference books.

Guide to U.S. Government Statistics (John L. Andriot, Arlington, VA: Documents Index, 1961-), indexes federal statistical sources and outlines the statistical output of federal agencies.

2. Abstracts and Indexes: Many abstracts and indexes of U.S. government publications act as guides and assist the researcher in finding government publications. The best tools are the two products published by the Congressional Information Service (CIS): *Congressional Information Service Index/ Abstracts* (Wash., D.C.: CIS, 1970-), and *American Statistics Index (ASI)* (Wash., D.C.: CIS, 1974-). *CIS Index/Abstracts* is a monthly updated, comprehensive index and abstracting service of Congressional documents, reports, committee prints, hearings and special publications. It has a subject index and is arranged by committee or organization for easy browsing. Since the most widely accessible, and sometimes most informative, reference materials are Congressional publications (specifically the annual hearings discussed later), *CIS Index/Abstracts* is the best single tool for the military researcher. *ASI Index* is a monthly abstracts and index service covering U.S. government publications which contain statistical information. While its coverage is not complete, it includes the executive branch agencies and Congressional committees, and is a valuable tool. Other abstracts and index services are of varying utility or cover specific subject areas or agency publications. Useful for research in the military and strategic field are the following:

Monthly Catalog of U.S. Government Publications (Wash., D.C.: GPO, 1895-), a monthly bibliography (with author, subject and title indexes) of government publications. The *Monthly Catalog* includes those publications which are placed in the depository library system and are printed by the GPO. This includes a valuable (though somewhat late) listing of Congressional publications, CIA and Joint Publications Research Service publications, treaties and other material of the State Department, and GAO reports. The *Monthly Catalog* is very weak on Defense Department publications, specifically many mentioned in this guide. It does, however, list National Defense University monographs and technical and training reports of the defense agencies and the military services.

The Federal Index (Cleveland, OH: Predicasts, Inc., 1976-), a monthly (with annual cumulative index) index of documents dealing with the federal government including the *Congressional Record, Federal Register, Weekly Compilation of Presidential Documents, Commerce Business Daily, Law Week* and the *Washington Post*. With its variety of sources, the *Federal Index* is a valuable timesaver and tool, particularly for the *Congressional Record* and the *Federal Register*.

The *Federal Register* and the *Congressional Record* are two valuable resources accompanied by monthly and biweekly indexes. A number of technical information indexes also exist and are discussed in Section IIIC4.

3. *Government Organization:* Several organizations and agencies are responsible for different aspects of national security affairs in the U.S. Government. The *U.S. Government Manual* (Office of the Federal Register, Wash., D.C.: GPO, 1935-) is the "official handbook of the federal government" and is the basic guide to the organization, personnel, purposes and programs of the federal agencies. Issued annually, it is the primary source for researching the general background of federal agencies. Its coverage of the Defense Department is good for the headquarters organization in Washington but very little data is given on field commands or military forces. It does not cover in detail the organization of the Congress or Congressional committees. The *Congressional Directory* (Wash., D.C.: GPO, 1809-) is the Congressional counterpart to the *Manual* and is the official annual directory of Congress with listings of Congressional members (with short biographies, committees and committee assignments, as well as other useful reference material on the Congress). Another valuable reference is the *Congressional Staff Directory* (Charles E. Brownson, ed., Wash., D.C., 1959-), an annual supplement to the *Congressional Directory* with staff listings including aides, committee and special office staff, and brief biographies. Both of the Congressional directories serve as excellent telephone directories.

Specialized organizational and reference manuals include three chart services. These are loose-leaf services providing updates of organization charts of federal agencies, including the agencies and commands of the Department of Defense. The best of these services is the *Federal Organization and Personnel Directory* (Wash., D.C.: Carroll Publishing Co., 1974?-), a two-volume service with very good coverage of the Department of Defense, particularly the research and development community. The other two services are the *U.S. Government Organization Chart Manual* (Columbus, OH: Symetics Group, Inc., 1973-) and *U.S. Military and Government Organization Chart Service* (La Jolla, CA: Organization Chart Service, 1971?-), which contain organization charts with personnel listings and telephone numbers.

Telephone directories can be valuable reference tools. The *Federal Executive Telephone Directory* (Wash., D.C.: Carroll Publishing Co., 1974?-) is a bimonthly commercial publication that provides good general coverage of the federal government. The official agency telephone directories of the State and Defense Departments are excellent tools also:

Department of Defense Telephone Directory (Dept. of Defense, Wash., D.C.: GPO), a triennial directory with personel listings available on subscription service from the GPO.

Department of State, AID, ACDA and OPIC Telephone Directory (Dept. of State, Wash., D.C.: GPO), an annual directory and personnel listing also available from the GPO.

The State Department has also compiled a handbook, *Foreign Affairs Research: A Directory of Government Resources* (Dept. of State, Wash., D.C., November 1977), which is an updated draft of a 1967 Office of External Research report outlining offices within the federal government that deal with all aspects of foreign affairs. Many more in-depth manuals of Defense Department activities exist and are discussed in Section IV, *The U.S. Military.*

4. The Freedom of Information Act: When dealing with the government in seeking data, files or publications that have not previously been released into the public domain, the Freedom of Information Act (FOIA) is an important tool for the researcher. The FOIA, passed by Congress in 1955, states that the government must make information available upon request or must prove (under the nine exemptions of the Act) why the information is not available. A researcher does not have to prove why he or she should have it. One planning to use the FOIA in requesting information should become thoroughly familiar with its provisions and procedures. A copy of the Act appears in the *U.S. Government Manual.* Two organizations can assist the researcher and provide more detailed information on how to use the Act:

Center for National Security Studies
122 Maryland Avenue, N.E.
Washington, D.C. 20002
(202) 544-5380

FOI Clearing House
2000 P Street, N.W.
Washington, D.C. 20036
(202) 785-3704

The FOIA is especially useful when dealing with the military because of the provision that the recipient of an FOIA request for information must respond within 10 working days. If you are unsuccessful in getting information from the public affairs offices, an FOIA request will often yield the desired information. (Many Public Affairs Officers are not familiar with what information is available, leaving the researcher at the mercy of either a good or bad Public Affairs Officer.) A researcher can request that the files or reports be made available for inspection, eliminating the cost of reproduction and part of the time lag.

TABLE 3
FREEDOM OF INFORMATION OFFICES
OF THE NATIONAL SECURITY COMMUNITY

Central Intelligence Agency
Information and Privacy
  Coordinator
Washington, D.C. 20505
(202) 351-5656

Department of Energy
Director, Div. of FOI and Privacy
  Acts
Forrestal Bldg., Room GA 152
1000 Independence Ave., S.W.
Washington, D.C.
(202) 252-6020

National Aeronautics and Space
  Administration
Public Information Service Chief
400 Maryland Ave., S.W.
Washington, D.C. 20546
(202) 755-8341

Department of State
Director, FOI Staff
2201 C St., N.W., Room 2815
Washington, D.C. 20520
(202) 632-0772

Arms Control and Disarmament
  Agency, Dept. of State
Freedom of Information Officer
320 21st St., N.W., Room 5534
Washington, D.C. 20451
(202) 632-0760

Agency for International
  Development, Dept. of State
Chair, Ofc. of PA, FOI and
  Privacy Acts
2201 C St., N.W., Room 2849
Washington, D.C. 20523
(202) 632-1850

Chief, Division of Information
  (Code PA)
HQMC, Room 1129
Washington, D.C. 20330
(202) 694-4308

Coast Guard, Department of
  Transportation
FOI Act Coordinator
400 7th St., S.W., Room 8324
Washington, D.C. 20590
(202) 426-2267

Office of the Assistant Secretary
  of Defense (Public Affairs)
Director, Freedom of Information
  and Security Review
Room 2C57, Pentagon
Washington, D.C. 20301
(202) 697-1160

Chief, Records Management
  Division
Office of the Adjutant General
Department of the Army
Attn: DAAG-AMR-S
Washington, D.C. 20314
(202) 693-1847

Chief, Naval Records
  Management Division
Department of the Navy
OP-09B1
Room 5E613, Pentagon
Washington, D.C. 20350
(202) 697-1459

Director, Defense Contract
  Audit Agency, CMA
Cameron Station
Alexandria, VA 22314
(703) 274-7285

Director, Defense Intelligence
  Agency
RTS-3A
Washington, D.C. 20301
(202) 692-5766

Director, Defense Investigative
  Service (D0020)
Room 2HO63, Forrestal Building
Washington, D.C. 20314
(202) 693-1740

Director, Defense Logistics
  Agency
DLA-XA
Cameron Station
Alexandria, VA 22314
(703) 274-6234

Director, Defense Mapping
  Agency
Naval Observatory, Building 56
Washington, D.C. 20305
(202) 254-4431

Director, Defense Nuclear Agency
Public Affairs Office
Washington, D.C. 20305
(202) 325-7095

Director, Defense
  Communications Agency,
  Code 105
Washington, D.C. 20305
(202) 692-2009

Director, National Security Agency
Attn: D4 FOIA Office
Fort George G. Meade, MD 20755
(301) 688-6964

Freedom of Information Manager
Department of the Air Force
HQ, USAF/DADF
Room 4A1088C, Pentagon
Washington, D.C. 20330
(202) 697-3467

Table 3 lists the major FOIA offices in the national security community. Every base and command has an FOIA officer, and requests may be directed to them. Obviously, much information within this field is classified, but the researcher should remember that many reports receive limited distribution within Washington because only five or ten copies are made and not because they are classified. The justification material discussed in Section IVC is a good example.

Another tool that the researcher should be familiar with is *Declassified Documents: Retrospective Collection* (Wash., D.C.: Carrollton Press, 1976-), a quarterly index and abstracts service. The *Declassified Documents Quarterly Catalog* (Part I; Catalog of Abstracts, Part II: Cumulative Subject Index) updates the *Retrospective Collection* with documents the U.S. government has declassified on the basis of FOIA requets, mandatory security reviews or requested reviews. This service can prove to be a tremendous source of background information for the researcher.

B. CONGRESSIONAL INFORMATION

Congressional documents probably provide the greatest source of information on military and strategic affairs—possibly even more than the Department of Defense—for three reasons: their regularity, accessibility and volume. Congressional publications are the most accessible to the researcher because they are well-indexed in the *CIS Index/Abstracts* and can be obtained without charge by writing one's Congressional representative or the committee of origin, or for a nominal fee from the GPO. They are also readily available in all government depository libraries and to subscribers of the CIS microfiche service.

Congressional publications are regularly produced with over 18 sets of publications of the annual major military hearings (see Table 5) and numerous annual reports and studies. The amount of material is tremendous, with the 18 sets of hearings totaling some 60 volumes and tens of thousands of pages. These annual hearings are not nearly the total output of the Congress, with many reports and studies being prepared by the various committees and subcommittees and by the research and oversight arms of Congress.

1. The Congressional Documents System: The use of Congressional documents requires an understanding of the organization and institutions of Congress, the types of publications it produces, and the role that Congress plays in the defense and foreign policy processes.

Congressional organization is outlined in two publications previously mentioned, the *Congressional Directory* and the *Congressional Staff Directory*. Standing committees of the House and Senate are responsible for oversight and legislation in functional areas. There are a number of committees and subcommittees whose primary responsibilities are defense and international affairs. Table 4 lists these primary committees and subcommittees. Each committee produces a quarterly *Legislative Calendar* which presents the status of bills and resolutions referred to the committee, as well as information on committee membership, rules, jurisdiction and publications. Another publication of the committee is the *Annual Report*, which differs greatly from committee to committee. The *Report of the Activities of the Committee on Armed Services, United States Senate* (Wash., D.C.: GPO, 1969-), for instance, summarizes the committee's legislative actions of the previous year on authorizations, oversight and review.

One doing current research on Congressional participation will be interested in using the tools which follow Congressional activity and publications. The various publications of Congress, the 15,000 or so bills per Congress, the hearings, committee prints, reports and documents, are all reported comprehensively in *CIS Index/Abstracts*, the *Congressional Record* and *CCH Congressional Index*. The *Congressional Record: Proceedings and Debates of the Congress* (Wash., D.C.: GPO, 1873-) is

TABLE 4
CONGRESSIONAL COMMITTEES AND SUBCOMMITTEES IN THE NATIONAL SECURITY FIELD

HOUSE: (all addresses are Washington, D.C. 20515)

Committee on Appropriations
Suite H218, The Capitol, (202) 225-2771
  Defense
  Foreign Operations
  Military Construction
  State, Justice, Commerce and Judiciary

Committee on Armed Services
Suite 2120, Rayburn House Office Building, (202) 225-4151
  Procurement and Military Nuclear Systems
  Seapower and Strategic and Critical Materials
  Research and Development
  Military Personnel
  Investigations
  Military Installations and Facilities
  Military Compensation
  Special Subcommittee on NATO Standardization,
    Interoperability and Readiness

Committee on the Budget
Suite 214, Cong. Ofc. Bldg. Annex No. 1, (202) 225-7200
  Defense and International Affairs

Foreign Affairs (formerly International Relations)
Suite 2170, Rayburn House Office Building, (202) 225-5021
  International Security and Scientific Affairs
  International Operations
  Europe and the Middle East
  Asian and Pacific Affairs
  International Economic Policy and Trade
  Inter-American Affairs
  Africa
  International Organizations

Permanent Select Committee on Intelligence
Room H405, The Capitol, (202) 225-4121

SENATE: (all addresses are Washington, D.C. 20510)

Committee on Appropriations
Suite 1235, Dirksen Senate Office Building
(202) 224-3471
  Defense

Military Construction
   Foreign Operations
   State, Justice, Commerce and the Judiciary

Committee on Armed Services
Suite 212, Russell Senate Office Building (202) 224-3871
   Arms Control
   General Procurement
   Manpower and Personnel
   Research and Development
   Military Construction and Stockpiles
   Procurement Policy and Reprogramming

Committee on the Budget
Room 208, Carroll Arms Annex, 301 First St., N.E., (202) 224-0642

Committee on Foreign Relations
Suite 4229, Dirksen Senate Office Building, (202) 224-4651
   International Economic Policy
   Arms Control, Oceans, International Operations
      and Environment
   African Affairs
   European Affairs
   East Asian and Pacific Affairs
   Near Eastern and South Asian Affairs
   Western Hemisphere Affairs

Select Committee on Intelligence
Room G-308, Dirksen Senate Office Building, (202) 224-1700
   Intelligence and the Rights of Americans
   Budget Authorization
   Collection and Production
   Charters and Guidelines

---

especially valuable to the researcher for its daily digest of bills and its "Extension of Remarks" section. It includes a subject and names index on a biweekly basis. *CCH Congressional Index* (Chicago: Commerce Clearing House, 1947-) is a weekly-updated index of public bills, with information on status, author and sponsors, and voting records. The *CQ Weekly Report* (Wash., D.C.: Congressional Quarterly, Inc., 1946-) is an excellent review of current Congressional activity. It includes much raw material, including statistical data, votes, Presidential messages, news conferences, etc. Congressional Quarterly also publishes the annual *CQ Almanac* and a set of weekly monographs called *Editorial Research Reports.*

Congressional involvement in national security affairs is of such importance that an understanding of its role as well as its products is essential. The role of Congress in the military and strategic fields has been discussed in a number of recent works, some of which may serve as resources:

Foreign Policy by Congress (Thomas M. Franck and Edward Weisband, New York: Oxford University Press, 1979).

A Responsible Congress: The Politics of National Security (Alton Frye, New York: McGraw Hill, 1975).

Congress, Information and Foreign Policy (Report prepared for the Committee on Foreign Relations, U.S. Senate, by the CRS, LC, Wash., D.C.: GPO, 1978).

Congressional Power: Implications for American Security Policy (Richard Haass, London: IISS (Adelphi Paper #153), 1979).

The U.S. Senate and Strategic Arms Policy, 1969-1977 (Alan Platt, Boulder, CO: Westview, 1978).

The Congressional Research Service's annual report, *Congress and Foreign Policy, 19__* (Report prepared by the CRS for the Committee on Foreign Affairs, U.S. House of Representatives, Wash., D.C.: GPO, 1974-), is a review of Congressional action in the foreign policy field.

2. *Defense and Foreign Policy Sources:* As stated above, the hearings (see Table 5) are the greatest single source on U.S. military and strategic affairs. The four sets of hearings held before the House and Senate Appropriations and Armed Services Committees and Subcommittees are comprehensive examinations of U.S. military posture. These annual authorizations and appropriations hearings consider military spending, the operations budgets of the various branches of the military, and various aspects of military posture and readiness. Normally, the following broad subjects are covered in detail:

Overview of U.S. Defense Posture
   Strategic Forces and the Strategic Balance
   General Purpose Forces and Regional Defense Balances
Current Posture of the Armed Services
Manpower and Training Programs and Strengths
Research, Development, Test and Evaluation
Procurement of Weapons
Reserve Forces

Published hearings contain not only the transcript of testimony and exchanges at the hearings, but exhibit reports,

TABLE 5
ANNUAL CONGRESSIONAL HEARINGS IN NATIONAL SECURITY AFFAIRS

House Appropriations Committee
    Dept. of Defense Appropriations for 19__(6-8 parts)
    Military Constructions Appropriations for 19__(4-5 parts)
    Foreign Assistance Appropriations for 19__(4-6 parts)

House Armed Services Committee:
    Hearings on Military Posture and Dept. of Defense Authorization Requests for FY__(6-7 parts)
    Military Construction Authorizations for Appropriations (Hearings on HR_____ to authorize certain construction at military installations) FY__
    Dept. of Energy National Security and Military Application of Nuclear Energy Authorization for 19 __

House Budget Committee:
    FY__ Overview of Budget for Defense and International Affairs (FY__ Defense Budget Overview)

House Foreign Affairs Committee:
    Foreign Assistance Legislation for FY__ (7-9 parts)

Senate Appropriations Committee:
    Dept. of Defense Appropriations FY__(6-7 parts)
    Military Construction Appropriations FY__
    Foreign Assistance and Related Programs Appropriations FY__(1-2 parts)

Senate Armed Services Committee:
    Dept. of Defense Authorizations for Appropriations for FY__ (Authorization of appropriations for military procurement, research and development, and active duty, selected reserve and civilian personnel strengths) (11 parts)
    Military Constructions Authorizations FY__
    FY__ Dept. of Energy Authorizations for Atomic Energy Defense Activities

Senate Budget Committee:
    First Concurrent Resolution on the Budget, FY__, Vol. I (Defense)

Senate Foreign Relations Committee:
    Foreign Relations Authorizations FY__
    International Development Assistance Authorizations FY__
    International Security Assistance Programs FY__

statistical data, articles and studies. The only index of the defense-related hearings is the annual *Congressional Testimony Index* (Greenwich, CT: DMS, Inc.). This index includes over 15,000 references to weapons systems mentioned in the House and Senate Armed Services and Appropriations hearings on the defense budget.

In addition to the major hearings on U.S. military posture, hearings are held on U.S. foreign aid, military assistance, and foreign relations by the House Foreign Affairs and Senate Foreign Relations Committees and Subcommittees annually. These hearings normally cover the following broad areas:

U.S. Security Supporting Assistance
U.S. Economic Assistance
U.S. Development Assistance
Human Rights Policy
U.S. Policy, Relationships and Aid in Regional Areas
  Africa
  Europe and the Middle East
  Latin America
  Asia

Other hearings held annually include those which examine proposed military construction programs of the coming year, hearings on nuclear weapons programs of the Department of Energy, and lesser hearings on the operations of the State Department, the Arms Control and Disarmament Agency, and other national security related agencies. Ad hoc hearings are held periodically (many are referenced later in this guide) on varied subjects: manpower programs (recruiting, the all-volunteer force, unionism), fraud and waste, NATO posture, stockpiling of strategic materials, etc.

Other Congressional publications include specially commissioned studies (see next section), analysis of federal legislation and activities, background and reference materials, working aids, and position papers. A detailed discussion of the Congressional calendar, hearings and reports appears in Section IVC, *The Defense Budget.*

3. *Congressional Agencies:* Four major Congressional agencies prepare reports for Congress, and a number of caucus groups and partisan committees examine military and strategic affairs. The publications of these organizations are announced and procured in various ways, mostly outside the GPO system.

The Congressional Budget Office (CBO) was established in

1974 to provide information on budget and fiscal programs, budget analysis and spending alternatives to Congressional committees and members of Congress. The National Security and International Affairs Division prepares specific reports, providing Congress with projections and alternatives regarding force structure and procurement decisions. These reports include *Budget Issue Papers, Background Papers* and *Staff Working Papers*. CBO publications are an excellent source of information for researchers. CBO analysis usually includes background on a given issue, the present status, and alternatives for the future. Single copies can be obtained from:

> Office of Intergovernmental Relations
> Congressional Budget Office
> House Office Building Annex No. 2
> Second and D Streets, S.W.
> Washington, D.C. 20515
> (202) 225-4616

Many CBO publications are sold by the GPO, and a list of all CBO publications to Fall 1980 (the *CBO List of Publications*) is available. Most CBO publications are referenced in the subject sections of this guide.

The General Accounting Office (GAO) and the Office of the Comptroller General provide studies and audit reports to Congress either on request or by legislative requirement. The International Division reviews governmental participation in assistance programs and foreign policy, security and defense. The Science and Technology Subdivision reviews defense and nuclear research and development, and science and technology programs. The GAO monitors the status of major weapons systems being acquired, assessing cost, development, production and deployment schedules. GAO and Comptroller General Reports are not available through the GPO. Single copies are available free from:

> U.S. General Accounting Office
> DHISF, Box 6015
> Gaithersburg, MD 20760
> (202) 275-6241

A semiannual list of GAO publications (*General Accounting Office Publications*) covers reports to Congress and federal agencies as well as GAO-delivered Congressional testimony. The *Monthly GAO Documents: Catalog of Reports, Decisions, and Opinions, Testimonies and Speeches* (GAO, Wash., D.C.: GPO, 1976-) contains a comprehensive subject and names index and abstracts of GAO publications. An annual report by the GAO, *Summaries of Conclusions and Recommendations on Department of Defense Operation* (Wash., D.C.: GAO), summarizes conclusions and recommendations resulting from audits conducted over the previous year.

The Office of Technology Assessment (OTA), established by Congress in 1972, analyzes the impact of technology and technological programs and identifies alternative technological methods and programs. The OTA conducts analysis and assessment of current research and development and provides this information to committees of Congress. A number of OTA studies would be of interest to the military researcher, including *The Effects of Nuclear War* (OTA, Wash., D.C.: GPO, 1979), which describes the effects of a nuclear war. The *Annual Report to the Congress by the OTA* (OTA, Wash., D.C.: GPO, 1973-) includes a list of OTA reports. A list of OTA publications is available from:

> Office of Technology Assessment
> U.S. Congress
> Washington, D.C. 20510
> (202) 224-8996

The Congressional Research Service (CRS), an arm of the Library of Congress, prepares many reports for Congress. By law, CRS produces analyses for the use of Congress and Congressional committees and does not have a system for providing its reports to the general public. CRS studies, however, have the reputation of being unbiased and sophisticated, making them valuable to the military and strategic affairs researcher. Many of the substantial studies undertaken by CRS for Congressional members or committees are published as Committee Prints or placed in the *Congressional Record*. One can trace these studies by using *CIS Index/Abstracts* or a semiannual CRS letter, "CRS Documents in the Public Domain," which cites reports which have been made public. Less substantial (and highly perishable) reports are published as CRS Issue Briefs. These Issue Briefs, distributed to Congressional members and committees, are not published but are available by writing to one's Senator or Representative. Many Issue Briefs are produced in the military and strategic fields, and they are updated as legislation or the world situation changes. Many recent Issue Briefs are referenced in later sections of this guide.

In addition to the four major organizations discussed above, two organizations of members of Congress do research on military and strategic affairs. These organizations produce reports and fact sheets, and give assistance to their members in many areas:

> Members of Congress for Peace Through Law
> Room 3538, House Annex No. 2
> Washington, D.C. 20515
> (202) 225-8550

> Democratic Study Group
> Room 1422, Longworth HOB
> Washington, D.C. 20515
> (202) 225-5858

C. EXECUTIVE BRANCH INFORMATION

Much information that does not appear in Congressional documents, as well as unpublished data, can be obtained from the agencies of the executive branch that deal with national security. The *U.S. Government Manual* describes the organization and functions of the agencies. Many areas of national security are the responsibility of more than one agency, making a thorough understanding of the agencies essential.

1. Presidential Information: The basic source of Presidential information is the *Weekly Compilation of Presidential Documents* (Office of the Federal Register, Wash., D.C.: GPO, 1965-), which publishes Executive Orders, press materials, announcements, addresses and appointments; and reviews in general terms the President's activities. The *Weekly Compilation* is indexed in the *Federal Index.* Very little information on military affairs comes from the President except in his speeches. The Office of Management and Budget is responsible for and publishes the federal budget and supplementary materials to the budget (these are described in Section IVC). The National Security Council advises the President and coordinates national security functions for the government. Most of the information that a researcher would be seeking from the government, however, would come from one of the three major national security agencies: the CIA, the State Department or the Defense

Department.

2. The Central Intelligence Agency: Since 1972 the CIA has been making a number of its unclassified publications of reference value available to the public. They are mostly working aids used by intelligence analysts. A number of the serials are referenced in this guide. A good article describing CIA publications is "The CIA, Its History, Organization, Functions and Publications" (Gregory A. Benadom and R.U. Goehlert, in *Government Publications Review*, Vol. VI, No. 3, 1979). A complete list of publications released is *CIA Publications Released to the Public Through Library of Congress: DOCEX Listing for 1972-1977* (CIA, National Foreign Assessment Center (NFAC-78-10002), Wash., D.C., July 1978), and a supplement "Unclassified CIA Publications Released Through DOCEX: Calendar Year 1978" (CIA, reproduced, n.d.).

Currently released publications of the CIA as of January 1, 1979 are available from the National Technical Information Service (NTIS) of the Department of Commerce (see Section IIIC4) either individually, by full subscription or by title subscription. Previously released publications are available for purchase from the:

Document Expediting Project (DOCEX)
Exchange and Gifts Div., Library of Congress
Washington, D.C. 20540
(202) 287-5253

The CIA publishes a series of maps which are available for purchase from the GPO. Two arms of the CIA, the Joint Publications Research Service (JPRS) and the Foreign Broadcast Information Service (FBIS), produce specialized reports which are available to the general public by subscription. The JPRS (1000 N. Glebe Rd., Arlington, VA 22201) translates and abstracts foreign language political and technical periodicals and media articles and makes them available in a series of reports available by subscription through NTIS. The subscription is a standing order for all reports issued in the year. Table 6 lists the various reports and the approximate number of issues per year. The price varies from $10-$450 depending upon the frequency of issue. An index to JPRS translations produced commercially, the Bell & Howell *Transdex Index* (Bell & Howell, Wooster, OH), includes a key word index, names index and bibliographic index.

The Foreign Broadcast Information Service (FBIS) pro-

TABLE 6
JOINT PUBLICATIONS RESEARCH SERVICE TRANSLATION SERIES

Translations on:
- North Korea (70/year)
- Japan (20/year)
- Vietnam (90/year)
- Mongolia (5/year)
- P.R.C.: Social, Economic, Military, Scientific and Technological Information (125/year)
- P.R.C.: Scientific Abstracts (18/year)
- P.R.C.: Agriculture (25/year)
- P.R.C.: Plant and Installation Data (15/year)
- P.R.C.: Biomedical and Behavioral Sciences (40/year)
- Latin America (170/year)
- Sub-Saharan Africa (170/year)
- Western Europe (170/year)
- Near East and North Africa (140/year)
- South and East Asia (20/year)
- Eastern Europe: Scientific Affairs (30/year)
- Eastern Europe: Political, Sociological and Military Affairs (150/year)
- Eastern Europe: Economic and Industrial Affairs (150/year)
- U.S.S.R.: Military Affairs (100/year)
- U.S.S.R.: Political and Sociological Affairs (100/year)
- U.S.S.R.: Economic Affairs (50/year)
- U.S.S.R.: Biomedical and Behavioral Sciences (40/year)
- U.S.S.R.: Resources (100/year)
- U.S.S.R.: Industrial Affairs (72/year)
- U.S.S.R.: Sociological Studies (4/year)
- U.S.S.R.: Trade and Services (80/year)
- U.S.S.R.: Science and Technology: Physical Sciences and Technology (35/year)
- U.S.S.R.: Space Biology and Aerospace Medicine (6/year)

Translations from "Kommunist" (18/year)
Translations from "Red Flag" (12/year)
Problems of the Far East (4/year)

---

duces a series of 8 daily reports of translated foreign news transmissions, editorials, and local media features. Like JPRS products, these are excellent sources of raw information for the analyst doing advanced research. The FBIS *Daily Reports* are available by subscription from NTIS at a cost of $150 annually (for one volume) and $50 for each additional volume. The eight volumes are:

    Vol. I: People's Republic of China
    Vol. II: Eastern Europe
    Vol. III: Soviet Union
    Vol. IV: Asia and Pacific
    Vol. V: Middle East and North Africa
    Vol. VI: Latin America
    Vol. VII: Western Europe
    Vol. VIII: Sub-Saharan Africa

Although a general index of FBIS reports does not exist, the *Daily Report: People's Republic of China* (Vol. I) has great research potential due to an index issued by Newsbank, Inc. (22 W. Putnam Ave., Greenwich, CT 06830). The subject and names index includes listings for personalities, institutions and organizations.

3. *The Department of State:* The State Department is responsible for the conduct of U.S. foreign policy, and publishes a great deal of material of interest to researchers, from news releases to a monthly magazine and a number of annual reports. The Bureau of Public Affairs coordinates most of these publications, coordinates inquiries from the public, and helps researchers contact functional experts within the Department who may be of additional assistance. The Office of Public Communications within the Bureau publishes the monthly *Department of State Bulletin, Special Reports, GISTS* (informal fact sheets on various foreign policy issues), *Selected Documents, Current Policy, Background Notes on the Countries of the World,* news releases, papers and pamphlets. Anyone can be placed on a mailing list for these publications by writing to:

Correspondence Management Division
Office of Public Communication
Bureau of Public Affairs
Department of State
Washington, D.C. 20520
(202) 632-1394

The Public Information Service answers most of the inquiries on foreign affairs directed to the executive branch and can be of assistance to researchers in answering policy questions dealing with U.S. foreign policy, security and other assistance programs, treaties and commitments and political-military affairs. The State Department encourages contact between qualified researchers and functional desk officers or analysts.

A researcher can either contact the desk officer or bureau public affairs representative directly or have the Public Information Service arrange an interview. The representatives are listed in the State Department telephone directory. Inquiries and requests for assistance should be directed to:

> Public Information Service
> Office of Public Communications
> Bureau of Public Affairs
> Department of State
> Washington, D.C. 20520
> (202) 632-6575

Many of the publications of the State Department and its annual reports are referenced in later sections of this guide. Testimony delivered by the Secretary of State and other high officials as well as major statements and speeches are available from the Bureau of Public Affairs. The *Department of State Bulletin* often contains excerpts and copies of these. Most of the publications of the State Department are sold by the GPO.

The Office of External Research of the Bureau of Intelligence and Research produces a series of bibliographic pamphlets that are especially valuable to area studies specialists. These pamphlets list unpublished specialized research papers undertaken for the State Department or placed in the Department's central reference library. The Office of External Research sponsors and contracts this research on a variety of subjects by outside experts and think tanks. A quarterly listing of newly-initiated research projects, *Government-sponsored Research on Foreign Affairs* (Department of State, Wash., D.C.: GPO), is available on subscription from the GPO. Other listings of papers published by the Office of External Research are *Foreign Affairs Research Papers Available*, a monthly accessions list with annual cumulations of papers by geographic regions; and *Foreign Affairs Special Papers Available*, which cover the following separate issues:

> Africa, Sub-Sahara
> American Republics
> East Asia and Pacific
> Europe and Canada
> Near East, South Asia and North Africa

People's Republic of China
U.S.S.R.

A good deal of information is available on the operations of the State Department and the actual conduct of foreign affairs. The *Department of State Newsletter* reports internal State Department developments and news, and personnel policy and assignments. The *Foreign Service Journal*, a commercial publication, reports on life and work in the Foreign Service. The official regulations of the State Department are contained in the *Foreign Affairs Manual*, a multi-volume set of internal regulations, organizations, and operations instructions. Annual Congressional hearings are held on the operations of the State Department and the conduct of foreign relations by the House and Senate Appropriations and Foreign Affairs/Relations Committees.

One doing advanced research on U.S. foreign relations may not be able to find the required material in the Library of Congress or other specialized libraries. The Department of State Library will admit a qualified researcher to use its collections if prior arrangements are made and the Department library does not have more pressing commitments. This can be arranged through the Public Information Service.

Two agencies of the State Department, the Agency for International Development (AID) and the Arms Control and Disarmament Agency (ACDA), have separate public affairs offices and prepare publications and other materials of interest to researchers. The Agency for International Development, besides being responsible for general U.S. development assistance programs, is also responsible for the execution of the U.S. "security supporting assistance" program. This program involves economic aid which the U.S. believes can be translated into political support to promote stability and manage crisis. These programs are examined in the annual foreign assistance hearings (see Table 5) and are also reported in AID publications mentioned later in this guide. Information on AID programs and operations are available from:

Office of Public Affairs
Agency for International Development
Department of State
Washington, D.C. 20523
(202) 632-1850

The Arms Control and Disarmament Agency is responsible

for the formulation and implementation of U.S. policies in the areas of arms control and disarmament, non-proliferation, and arms trade. It produces a number of annual publications referenced in this guide. The operations and organization of ACDA are examined annually in hearings before the House Foreign Affairs and Senate Foreign Relations Committees. The Office of Public Affairs can provide assistance to researchers and facilitate meetings between qualified researchers and ACDA staff officers. It can also answer questions about ACDA operations and policy questions dealing with areas under ACDA jurisdiction.

Office of Public Affairs
U.S. Arms Control and Disarmament Agency
320 21st St., N.W.
Washington, D.C. 20451
(202) 632-9504

4. The Department of Defense: The Defense Department is a source of information which many researchers often bypass. The belief that the Defense Department is more secretive or less forthcoming than other government national security agencies is very pervasive. This view is not necessarily true; yet there are many reasons why one is not always successful when seeking information from the DOD. One problem unique to the Defense Department is its size. Researchers seek not only information about military policy in many different fields, but information about weapons systems, military forces, management, contracts, internal procedures, expenditures and operations. The mere size of the Defense Department defies complete comprehension and clear parameters. It is folly to believe, therefore, that any one public affairs officer knows everything. In fact, the military system breeds specialists, and public affairs officers are often nothing more than facilitators who act as buffers between the media, the public and the specialists. Another problem related to size is organization. The DOD is a massive organization made up of many independent operational components. One must remember that the military system is very hierarchical. The organization on the top makes policy which is passed down to lower organizations which implement it, and which pass it down to lower organizations which further implement it and so forth. One must therefore understand DOD organization and who is authorized to speak about what policy and who is not. Half the problem of getting information from the Defense Department is getting the request to the right person or office in the first place. However, with the mass of data and information produced, the Freedom of Information Act (FOIA) and persistence, one should be relatively successful in satisfying his or her research needs. Finally, in talking about the Defense Department, one needs to talk about spirit. How hard an official in the Department pursues information in response to a request is dependent on a human equation. The work load of the official, the researcher's importance or *bona fides,* the specificity of the request, the urgency of the tone, are all factors which will influence service. Be clear, specific, persistent, and if all else fails, use the FOIA. It is your right.

The first step in learning about the Defense Department is understanding its organization. This is discussed in Section IVA. Each of the components of the Defense Department has a separate public affairs, publications, regulations and management system. The common factors to keep in mind are outlined below. Understanding what type of information is already published and in the public domain, and what type is not public but is generally obtainable, is essential.

Public Affairs Information: Each public affairs component of the Defense Department operates differently. Generally, the Public Affairs Office (PAO) produces fact sheets and news releases, answers telephonic and written requests, and assists in finding material to satisfy research questions. However, news media requests and keeping DOD's internal information system going demand most of the PAO's time, leaving less time and energy for researchers and members of the general public. The responsibility for dealing with these groups is often delegated to a "Community Relations" staff or public correspondence desk officer who takes charge of answering requests from both third-grade students and advanced researchers. The establishment of credentials when requesting information is essential. State whether you are working on a commissioned piece that will be published, at what level you are conducting your research, and your official affiliation.

Internal Information: The Defense Department internal information system is used to keep the members of the armed services and the employees and affiliates of the military appraised of DOD developments and policies. This supports the thousands of base and organization newspapers and newsletters, the news services, the training establishments, and the individual soldiers and potential soldiers. There are about 100 periodicals of interest to researchers, ranging from scholarly journals to general news magazines. Many of these periodicals are available by subscription from the GPO.

Regulatory and Doctrinal Information: Regulations govern each part of DOD's organization, operations and administration. This regulatory material is of great value because it sets out policy and procedures and gives background information on every aspect of the military system. The operational doctrine of the Armed Forces is also contained in regulatory publications. These doctrinal publications describe operational doctrine, tactics, and strategy and provide background information on the conduct of war and military operations.

Policy Information: Most of the policy information published by the Department of Defense is in the form of annual or periodic reports, testimony and justification material delivered to the Congress, and publications which review activities, expenditures and posture of the various components and activities of the military. These publications provide the researcher with much raw data and review many aspects of U.S. military policy and posture. Most of these publications are produced in response to legislative requirements and are directed at the Congress and other branches of the government. The most useful and common ones are referenced later in this guide.

Technical Information: Technical reports prepared by the research and development components of the Department of Defense or by private companies which perform research under contract for the DOD are valuable sources of information. Since the DOD tries to make these studies and reports accessible to the R&D community and to defense industries, they are well-indexed and controlled. This class of publications is not limited to technical reports of a scientific nature but includes student papers prepared at military schools and studies in the areas of military science, doctrine and tactics, the military balance, etc. Section IIID4 discusses the different types of technical reports and the methods of getting these reports.

(a) DOD Information: The Office of the Assistant Secretary of Defense (Public Affairs), also referred to as the DOD Public Affairs Office, is the central and highest focal point for the release of information to the public concerning the Department of Defense, the Joint Chiefs of Staff, and U.S. military and defense policy. Inquiries dealing with defense policy, the defense budget, or other areas under central management and jurisdiction of the Secretary of Defense and the Joint Chiefs of Staff should be directed to the DOD Public Affairs Office. The DOD Public Affairs Office is organized into a number of divisions which serve writers and researchers of different types.

The Staff Assistant for Public Correspondence answers the

mail directed to the DOD from the general public. It can provide publications, speeches and other materials, and respond to inquiries about defense policies and programs. One who is not a member of the media, a magazine writer or preparing a book should direct his or her requests to this office.

> Staff Assistant for Public Correspondence
> Office of the Assistant Secretary of Defense
>   (Public Affairs)
> Washington, D.C. 20301
> (202) 697-5737

The News Division handles requests from the regular media and freelance writers for information, interviews and background data. The News Division does not normally operate by answering written requests since it deals with people who are preparing material for immediate publication. It also prepares DOD news releases (about 3-5 a day). Desk officers within the division can answer telephonic requests and can be reached at (202) 697-5131.

The Audio-Visual Division handles requests from book and magazine writers who require assistance from the DOD in terms of interviews, visits to bases or pictorial support. This division can also refer researchers to the appropriate offices and assist in a fact-finding trip to Washington.

> Audio-Visual Division
> Office of the Assistant Secretary of Defense (Public
>   Affairs)
> Washington, D.C. 20301
> (202) 697-1252

The Office of the Assistant Secretary of Defense (Public Affairs) also has a Community Relations Division which handles liaison with organizations and associations and arranges speakers.

Defense Department publications and periodicals provide a general overview of defense policy and are of interest to any researcher in the military and strategic fields.

News releases and fact sheets: News releases are prepared to announce contract awards, major reorganizations or structural changes, ship commissioning, general officer appointments, military personnel strengths and base closures and realignments. Speeches by the Secretary of Defense and other high officials in the DOD are printed, but not as news releases. No fact sheets are produced by the DOD but requests for statistical data are often responded to with unpublished data or excerpts from Defense Department reports.

Periodicals: Not as many periodicals are produced at the Defense Department level or by Defense Department activities as by the armed services. Some periodicals of interest published by DOD-level or joint agencies and components are:

Asia Pacific Defense Forum
Defense/81 (formerly *Command Policy*)
Defense Management Journal
Defense Systems Management Review
DISAM Newsletter
Joint Perspectives
Military Law Review
Program Manager

Publications: Some frequently referenced Defense Department publications provide the basic documents on current defense programs and policy. Six publications provide most of the information that researchers require. These basic documents are:

Department of Defense Annual Report ("The Secretary's Posture Statement")
United States Military Posture for Fiscal Year 19__ ("The Joint Chiefs of Staff Military Posture Statement")
Program Acquisition Costs by Weapons System: DOD Budget, FY__
The FY__ Department of Defense Program for Research, Development and Acquisition
Manpower Requirements Report for FY__
Foreign Military Sales and Military Assistance Facts

This is by no means the extent of the recurring publications of use to the researcher. Most publications of the Defense Department are available to the public by writing to the Staff Assistant for Public Correspondence. There is a charge for some publications and some are only available from the originating office of the National Technical Information Service. The DOD Washington Headquarters Services Directorate for Information Operations and Reports (The Pentagon, Room 1C535, Wash., D.C. 20301) is the central distributor of many DOD reports dealing with logistics, personnel, financial management, and contracts. The *Catalog of DIOR Reports* lists 23 recurring publications and services compiled by the DOD and for sale by the Washington Headquarters Services.

Regulations: DOD regulatory material that sets policy is called Directives and Instructions. Other regulatory material which dictates administrative procedure and explains systems, organizations and operations are Handbooks, Manuals and Pamphlets. These five types of regulations—*DOD Directives, DOD Instructions, DOD Handbooks, DOD Manuals* and *DOD Pamphlets*—and the same administrative publications of the DOD agencies, joint activities and Unified Commands (a list of these appears in Table 13), are of tremendous use to researchers. *DOD Directives* and *Instructions* are indexed in the quarterly *DOD Directives System Quarterly Index* and *Quarterly Index of Final Opinions, Statements of Policy, and Administrative Staff Manuals and Instructions which Affect the Public* (DOD 5025.1) (Wash., D.C.: DOD). This publication, as well as copies of *DOD Instructions* and *Directives*, is available free (up to 5 items per request are allowed) from:

U.S. Naval Publications and Forms Center
5801 Tabor Avenue, Attn: Code 301
Philadelphia, PA 19120

DOD Handbooks, Manuals and *Pamphlets* are indexed in *Index of Administrative Publications* (DA Pam 310-1) and *DOD, JCS & Interservice Publications and Air Force Acquisition Documents* (AFR 0-4). They are available, most for a fee, from the same address.

Each of the defense agencies and the Unified Commands has similar regulatory material. If one is doing research into the administration, organization and operations of one of these organizations, the index of regulations or the appropriate regulatory material should be requested of the agency.

The Defense Intelligence Agency, like the CIA, makes many of its intelligence working aids and studies available to the public. A list of the publications available for purchase can be obtained from:

Defense Intelligence Agency
RTS-2A
Washington, D.C. 20301

(b) Air Force Information: The Air Force has a centralized public affairs system for dealing with information requests, from writers, researchers, and the general public. There are basically three points of contact. General, brief questions from non-media individuals, institutions and associations about organizations, equipment and programs which could be answered by prepared fact sheets, policy statements or other prepared material (the Air Force produces more of this type of material than any other component of the military) should be directed to:

>Department of the Air Force
>Office of Public Affairs (SAF/PAC)
>Community Relations Division
>Washington, D.C. 20330
>(202) 697-1128

The Community Relations Division will respond to a specific written request or telephone query but is not staffed or organized to do research. Questions about the Air Force for a book or magazine article should be directed to:

>Air Force Office of Public Affairs (AFOPA/MB)
>Magazines and Book Division
>1221 S. Fern St.
>Arlington, VA 22202
>(703) 695-5382

The Magazine and Book Division will assist a researcher in getting data from various organizations in the Air Force, set up interviews, obtain pictorial support, and clear access to the key historical depositories.

Writers working on pieces for weekly or daily media should contact the:

>Department of the Air Force
>Office of Public Affairs (SAF/PA)
>Media Relations Division
>Washington, D.C. 20330
>(202) 695-5554

These Air Force public affairs offices at the HQ, USAF level are generally able to answer inquiries relating to Air Force policy, forces, weapons, programs and organization. Researchers can also contact each Air Force major command (see Table 14) and installation, whose public affairs officer can be of assistance on more specific requests.

The majority of Air Force policy-related publications are copies of testimony delivered by officials of the Air Force in support of the Air Force portion of the defense budget. No formal Air Force publications are published regularly that are similar to the publications mentioned under DOD. Speeches and testimony of high Air Force officials are available from the public affairs office. Since the Air Force makes much more extensive use of fact sheets than any of the other services, much prepared information is often available in response to general research questions.

Many periodicals published by the Air Force (see Table 7) are of general interest, in addition to the hundreds of base and command periodicals and newspapers. Some of these publications are available from the GPO (see Appendix A) by subscription.

Air Force regulations and regulatory material are centrally published and available for purchase from:

Air Force Publications Distribution Center
2800 Eastern Boulevard
Baltimore, MD 20402
(301) 962-7282

Three major indexes to this material are of interest to researchers:

Guide to Indexes, Catalogs, and Lists of Department Publications (AFR 0-1) (Wash., D.C.: USAF).

Numerical Index of Standard and Recurring Air Force Publications (AFR 0-2) (Wash., D.C.: USAF), the most useful index with *Air Force Regulations* (AFR), *Air Force Manuals* (AFM), and *Air Force Pamphlets* (AFP).

Department of Defense, Joint Chiefs of Staff & Interservice Publications and Air Force Acquisition Documents (AFR 0-4) (Wash., D.C.: USAF).

Each of the Air Force major commands also publishes its own regulations, pamphlets and manuals which are indexed in the command regulations 0-2 (e.g., Tactical Air Command index of regulations is TACR 0-2). These regulations may be of interest for more advanced research.

(c) Army Information: The Army has a single office that is the point of contact for requests for information:
H.Q. Department of the Army
Office of the Chief, Public Affairs
Washington, D.C. 20310
(202) 697-5789

TABLE 7
PERIODICALS PUBLISHED BY THE AIR FORCE

Accounting and Finance Technical Digest
Aerospace Safety Magazine
Air Force Administrator
Air Force Comptroller
Air Force Engineering and Services Quarterly
Air Force Law Review
Air Force Policy Letter for Commanders
Air Force Policy Letter Supplement
Air Reservist
Air University Review
Airman
Combat Crew
Contracting and Acquisition Newsletter
Driver
The MAC Flyer
Maintenance
Medical Service Digest
Security Police Digest
Systems Management Newsletter
Talon
TIG Brief
Transportation Brief
USAF Fighter Weapons Review

    This office responds to requests from the news media, freelance writers, book and magazine writers. Some letters from the general public are routed to the Adjutant General's Office but requests for current information on policy, programs, organizations and weapons should be directed to the above office.

    The public affairs office also makes available testimony, speeches and policy announcements of the Secretary of the Army and other high officials of the Army. The testimony of the Secretary and of the Chief of Staff is bound into the annual "posture statement," which is also available.

    The periodicals published by the Army (see Table 8) cover many varied subjects of interest to the researcher. Of particular interest are the publications of the various branch schools of the Army, probably the most useful group of periodicals published by any of the branches of the military.

    Army regulations and regulatory material are published and distributed and are available for purchase from:

TABLE 8
PERIODICALS PUBLISHED BY THE ARMY

The Advocate
Air Defense Magazine
All Volunteer
Armor
Army Administrator
The Army Communicator
The Army Lawyer
Army Logistician
Army R,D,A
Army Reserve Magazine
Assembly
Commanders Call
The Engineer
Eurarmy Magazine
Field Artillery Journal
Infantry
The Journal of the USAINSCOM
Military Intelligence
Military Law Review
Military Media Review
Military Police Law Enforcement Journal
Resource Management Journal
Soldiers
Translog
USA Aviation Digest
USA Recruiting and Re-enlistment Journal
USA Security Assistance Bulletin

U.S. Army AG Publications Center
2800 Eastern Boulevard
Baltimore, MD 21220

There are five major indexes of publications. The various regulatory and training publications of the Army (see Table 9) are listed in these indexes:

Index of Administrative Publications (Regulations, Circulars, Pamphlets, Posters, Joint Chiefs of Staff Publications,

TABLE 9
ARMY REGULATORY DOCUMENTS AND PUBLICATIONS (SHORT AND LONG TITLES)

AR	Army Regulation
SubjScd	Army Subject Schedule
ATP	Army Training Programs
ATT	Army Training Test
DA Circ	Department of the Army Circular
DA Pam	Department of the Army Pamphlet
DA Poster	Department of the Army Poster
FM	Field Manual
FT	Firing Table
GO	General Order
GTA	Graphic Training Aid
LO	Lubrication Order
ROTCM	Reserve Officer's Training Corps Manual
SB	Supply Bulletin
SC	Supply Catalog
SM	Supply Manual
TA	Table of Allowances
TB	Technical Bulletin
TC	Training Circular
TD	Table of Distribution
TDA	Table of Distribution and Allowances
TJC	Trajectory Chart
TM	Technical Manual
TOE	Table of Organization and Equipment

DOD and Miscellaneous) (DA Pam 310-1) Wash., D.C.: USA), the most useful index of policy material.

Index of Doctrinal, Training and Organizational Publications (DA Pam 310-3) (Wash., D.C.: USA).

Index of Technical Manuals, Technical Bulletins, Supply Manuals (Types 7, 8 and 9), Supply Bulletins and Lubrication Orders (DA Pam 310-4) (Wash., D.C.: USA).

Index of Supply Catalogs and Supply Manuals (Excluding Types 7, 8, and 9) (DA Pam 310-6) Wash., D.C.: USA).

"How to Fight" Literature (DA Circ. 310-1) (Wash., D.C.: USA), a listing of new field manuals and other doctrinal and training literature.

The Army Major Commands (see Table 16) also publishes

regulations, manuals and pamphlets. These are indexed in the command 310-1 series pamphlet.

(d) Navy Information: The Navy has two primary contacts for information within its Office of Information: the Public Inquiries Branch and the Media Services Branch. The Public Inquiries Branch receives and answers letters sent to the Navy from the general public. It generally can answer brief questions or requests for information and publications about policy, weapons systems, organizations, programs and personnel. Often requests are referred to functional offices or commands in the Washington area.

> Department of the Navy
> Office of Information
> Public Inquiries Branch
> Washington, D.C. 20350
> (202) 695-0965

The Media Services Branch is responsible for dealing with requests for information from magazine writers, book writers and freelance journalists. These requests are normally for specific data needed for research or writing assignments. The Media Services Branch can arrange interviews and provide policy material and data to support research and writing projects.

> Department of the Navy
> Office of Information
> Media Services Branch
> Washington, D.C. 20350
> (202) 695-0293

No policy publications are published by the Navy other than the testimony and statements of the Secretary of the Navy, the Chief of Naval Operations and other high naval officials. The annual "posture statement" of the Chief of Naval Operations is bound and is available from the Office of Information.

Numerous periodicals published by the Navy (see Table 10) are of interest to the researcher. Many are available by subscription (see Appendix A) from the GPO.

Navy regulatory material is less centralized than that of the other services and this often proves very confusing. Each of the Naval commands and major offices of Navy headquarters issues regulations, all of which carry the same weight. In

TABLE 10
PERIODICALS PUBLISHED BY THE NAVY

Advisor
All Hands
Approach
Campus
Direction
European Scientific News
Fathom
The JAG Journal
The Log
Mech
Naval Aviation News
Naval Research Logistics
   Quarterly
Naval Research Reviews
Naval Reservist News
Naval War College Review
Navy Civil Engineer
The Navy Human Resource
   Journal
Navy Lifeline
Navy Policy Briefs
Navy Supply Corps
   Newsletter
Report of NRL Progress
Sealift
Surface Warfare Magazine
U.S. Navy Medicine
U.S. Navy Security
   Assistance Newsletter

addition, Navy regulations are called Instructions and Navy training and doctrinal material is published by three separate organizations. Table 11 lists the long and short titles of the most common Naval regulatory and training materials. The central office for publication and distribution of Navy regulatory material is:

    U.S. Naval Publications and Forms Center
    5801 Tabor Avenue
    Philadelphia, PA 19120

Most of the material listed below (unless specified otherwise)

TABLE 11
NAVY REGULATORY DOCUMENTS AND PUBLICATIONS (SHORT AND LONG TITLES)

Instructions:
AO Inst	Navy Administrative Office
BUMED Inst	Bureau of Medicine and Surgery
BUPERS Inst	Bureau of Naval Personnel
CHINFO Inst	Office of the Chief of Information
CNET Inst	Chief of Naval Education and Training
JAG Inst	Judge Advocate General
NAVAIR Inst	Naval Air Systems Command
NAVCOMPT Inst	Office of the Comptroller of the Navy
NAVELEX Inst	Naval Electronics Systems Command
NAVFAC Inst	Naval Facilities Engineering Command
NAVINT Inst	Naval Intelligence Command
NAVMAT Inst	Naval Materiel Command
NAVORD Inst	Naval Ordnance Systems Command
NAVPERS Inst	Bureau of Naval Personnel/Naval Military Personnel Command
NAVPUB Inst	Naval Publications and Printing Service
NAVSEA Inst	Naval Sea Systems Command
NAVSHIPS Inst	Naval Ships System Command
NAVSUP Inst	Naval Supply Systems Command
NAVTEL Inst	Naval Telecommunications Command
OCEANAV Inst	Office of the Oceanographer of the Navy
OCMM Inst	Office of Civilian Manpower Management
ONR Inst	Office of Naval Research
OPNAV Inst	Chief of Naval Operations
SECNAV Inst	Secretary of the Navy
WEASERV Inst	Naval Weather Service

COMTAC Publications: Communications and Tactical Warfare Publications):
ACP	Allied Communications Publications
ATP	Allied Tactical Publications
AXP	Allied Exercise Publications
FXP	Fleet Exercise Publications
NWIP	Naval Warfare Information Publications (being incorporated into NWPs)
NWP	Naval Warfare Publications

Other Publications:
NAVEDTRA P	Naval Education and Training Publication
NAVEXOS P	Executive Office of the Secretary of the Navy Publication

NAVMC   Naval/Marine Corps Publication
NAVSO P   Department of the Navy Staff Office Publication
NAVTRA P   Naval Training Publication

---

is available for purchase from the above address. The various indexes of Navy regulatory material are:

Consolidated Subject Index of Instructions by Washington Headquarters Organizations (NAVPUB-NOTE 5215) (Wash., D.C.: Navy, NPPS Management Office), lists the instructions of the commands in the Washington area.

Index of Forms and Publications (NAVSO P-2345) (Wash., D.C.: Navy, Ofc. of the Comptroller), publications of the Secretary of the Navy and Department of the Navy staff offices.

Naval Warfare Publications Guide (NWP-O) (Wash., D.C.: Navy, CNO), an index of Naval COMTAC (Communications and Tactical Warfare) publications and Allied publications.

List of Training Manuals and Correspondence Courses (NAVEDTRA 10061) (Wash., D.C.: Navy, CNET), a semiannual list of Naval training manuals.

Naval instructions and COMTAC publications are available from the Publications and Forms Center. Training publications are available from the:

Naval Training Equipment Center
Orlando, FL 32813

Chief of Naval Education and Training
NAS Pensacola, FL 32508

(e) Marine Corps Information: The Marine Corps has one point of contact for public affairs requests. The Office of the Director of Public Affairs (formerly the Office of the Director of Information) can answer letters, assist researchers and provide Marine Corps publications and background material on Marine Corps forces, equipment, policy and operations.

HQMC
Director of Public Affairs
Washington, D.C. 20380
(202) 694-1492/93

The Office of Public Affairs issues 1-2 news releases a day but most of its products are directed towards internal information. The Marine Corps does not publish periodicals like the other services, relying instead on the commercial publications of the Marine Corps Association: the *Marine Corps Gazette* and *Leatherneck*.

Marine Corps regulatory material is published and distributed at Marine Corps Headquarters in Washington and is available for sale from:

> Commandant of the Marine Corps
> HQMC (HQ-SP)
> Washington, D.C. 20380

Regulations covering policy are either *Marine Corps Orders* (MCOs) or *Marine Corps Bulletins* (MCBs). *Marine Corps Bulletins* are circulars which expire upon a certain date (normally in effect for 6 months or less). Doctrinal material includes *Fleet Marine Force Manuals* (FMFMs), *Naval/Marine Corps Publications* (NAVMCs), and *Landing Force Manuals* (LFMs). These publications are indexed in the following:

> *Marine Corps Directives System Semi-Annual Checklist* (MCB 5215) (Wash., D.C.: USMC), indexing MCOs and MCBs.
> *Listing, Status and Review Assignment Schedule of Marine Corps and Joint Doctrinal Publications* (MCB 5600) (Wash., D.C.: USMC), indexing LFMs, NAVMCs and FMFMs.

Two additional publication types not published by HQMC but of tremendous use are *Education Center Publications* (ECPs) and *Operations Handbooks* (OHs). These are published and are available from:

> Marine Corps Development and Education
>   Command
> Quantico, VA 22134

These publications are utilized for training texts and background materials at the Education Command and are more detailed and up-to-date sources in many instances than FMFMs. They contain more operational and organizational information than is normally in the official doctrinal series.

(f) Technical Information: Technical information and reports are a tremendous source of information for the researcher. The immense outpouring of government-produced and sponsored studies are put into a separate system for distribution and control from other governmental publications and regulatory materials, in order to make this material widely available to defense industry and to the R&D community. But technical reports and information are not limited to scientific and design studies dealing with weapons systems and technology. Many government publications of limited interest and many specially commissioned studies dealing with military sciences, forces, operations and doctrine, as well as many papers of officers at advanced DOD schools, are placed into the technical information system. There are two indexes of reports available to the public that the researcher should be familiar with.

The government-wide clearinghouse of technical information is the National Technical Information Service (NTIS) of the Department of Commerce. NTIS collects, announces and disseminates federal publications, data, scientific and technological information. This includes U.S. (and some cooperating foreign government) sponsored research, development and engineering reports, and many commissioned and contracted studies of the federal agencies. NTIS is also the major source of foreign language translated scientific information, primarily Russian, Chinese and Japanese originated material. NTIS receives the majority of its offerings in the military field from material releasable to the public (both unclassified and non-proprietary information) from the Defense Technical Information Center (formerly the Defense Documentation Center), an in-house DOD technical information system. The unclassified/unlimited reports and publications of the DOD and the research, development, test and evaluation (RDT&E) community are announced in the NTIS-produced *Government Reports Announcements/Index* (Springfield, VA: NTIS), a biweekly title listing organized into subject fields (including military sciences, history, law, political science and ordnance as well as a number of strictly technical fields). The companion Index issue contains a subject, corporate and personal author index. Copies of reports listed in the NTIS GRA/GRI are available for purchase from:

National Technical Information Service
Springfield, VA 22161
(703) 537-4660

NTIS also has a number of special services and will conduct searches and compile bibliographies for a fee. Many of these services, including subscription services (e.g., for the CIA, JPRS and FBIS products) are discussed in an NTIS pamphlet, "NTIS

Information Services: General Catalog No. 6" (NTIS, Jan. 1979) available from NTIS.

The other major data base of military related technical information is the *Annual DOD Bibliography of Logistics Studies and Related Documents, Jan. 19_* (Defense Logistics Studies Information Exchanges, Ft. Lee, VA: USA Logistics Management Center), with three quarterly supplements. This index and abstract lists reports and contract studies done for all DOD components (both internally produced studies and contracted studies) relating to defense industry, R&D, procurement, management, basing, personnel training and resources. Each issue includes a subject, contractor and contractee index. This product is particularly useful for advanced research relating to logistics and also contains abstracts of audit reports prepared by the audit agencies of the DOD and the armed services.

Another type of technical information is reports prepared by students at DOD schools. Some of these reports are indexed by GRA/GRI, but many are not. Table 12 lists the major DOD advanced schools which produce student reports of interest to researchers. Each of the schools produces a summary of its research and papers. The Air University index, *Air University*

TABLE 12
HIGHER MILITARY EDUCATIONAL INSTITUTIONS OF THE DOD

Air University, Maxwell AFB, Montgomery, AL 36112
    Air War College
    Air Command and Staff College
Armed Forces Staff College, Norfolk, VA 23511
Army Command & General Staff College, Ft. Leavenworth, KS 66027
Army War College, Carlisle Barracks, PA 17013
Defense Systems Management College, Ft. Belvoir, VA 22060
Inter-American Defense College, Ft. Lesley J. McNair, Washington, D.C. 20315
Marine Corps Command and Staff College, MCB Quantico, VA 22134
National Defense University, Ft. Lesley J. McNair, Washington, D.C. 20319
    National War College
    Industrial College of the Armed Forces
Naval Postgraduate School, Monterey, CA 93940
Naval War College, Newport, RI 02840

Abstracts of Research Reports (AU Lib., Maxwell AFB, AL: USAF, 1957-), is the most comprehensive for any school. It includes a list of faculty and staff studies and student papers prepared by the various schools of the Air University including the Air War College, Air Command and Staff College, and the Air Force Institute of Technology. The National Defense University also produces a series of "blue books," specially prepared texts for students in its correspondence courses, which are excellent sources of background information.

Finally, some of the larger think tanks produce their own indexes of reports, most of which are prepared with government money. The most widely known is *Selected RAND Abstracts: A Quarterly Guide to Publications of the Rand Corporation* (Santa Monica, CA: Rand, 1964-), which is a subject and author index (with abstracts) of reports, translations, and books covering current military affairs, arms control, decisionmaking, foreign policy, military manpower, conflict and terrorism. The Center for Naval Analysis produces a useful index, *Index of Selected Publications through Dec. 19__* (Alexandria, VA: CNA), which contains abstracts of papers by CNA staff members, either published by CNA or commercially. Other organizations which produce indexes and listings include the Hudson Institution, Institute for Defense Analysis, and Brookings Institution.

IV.
THE U.S. MILITARY

More information is available on the U.S. military than on any other military force in the world. From the mountain of government documents, Congressional hearings and publications, and the extensive military periodical field, researchers can construct a detailed picture of the U.S. military, its force structures, weapons, policies and practices. Understanding the current policies and issues of the U.S. military requires a grasp of the organization, programs, components, budget and activities of the Department of Defense (DOD) and the four Armed Services (Army, Navy, Air Force, and Marine Corps). This fourth part introduces the basic sources of information on the U.S. military and covers various areas of interest to researchers. It presents both general introductory and specialized data sources that can be used to formulate independent assessments and analyses of current issues facing the military.

A. DEPARTMENT OF DEFENSE ORGANIZATION AND BACKGROUND

Basic to an understanding of the military is appreciation of military organization and structure. A few good sources provide an introductory view of the overall organization of the Defense Department, and should be the first sources consulted for those not familiar with the military system. Very accessible is the *U.S. Government Manual,* a basic source available in almost every library. The manual outlines DOD organization, lists some of the most important officials, and explains the responsibilities of the three military services. Another good overview is provided in a recent commercial publication, *The U.S. War Machine: An Encyclopedia of American Military Equipment and Strategy* (James E. Dornan, Jr. (consultant), New York: Crown, 1978), which contains photographs, charts, tables and articles providing a good introductory picture of defense organization, the armed forces and weapons. A third source is *Defense Organization and Management* (Theodore W. Bauer and Eston T. White, Wash., D.C.: NDU, 1978), one of the volumes in the National Defense University (NDU) "blue book" series. This gives a good organizational overview of the various components of the DOD. Telephone books and organizational chart services are other good sources for understanding defense organization. Although they contain no explanatory material, they can be quite instructive. Some recommended sources include:

> Department of Defense Telephone Directory
> Federal Executive Telephone Directory
> Federal Organization and Personnel Directory
> Global Autovon Telephone Directory

1. DOD and DOD Agencies: A number of sources explain higher DOD organization, including the organization of the Office of the Secretary of Defense, the Joint Chiefs of Staff, the defense agencies, the Unified Commands, and the various joint and defense activities. A brief description of the components of the DOD, and its responsibilities, offices and commands, is *U.S. Department of Defense Fact Sheet, 1978* (Wash., D.C.: OASD (PA), 1978), a pamphlet issued by the DOD Public Affairs Office. A similar guide which is kept more up-to-date is "DOD Brief of the Organization and Functions: Secretary of Defense, Deputy Secretary of Defense, Defense Staff Offices, Organization of the Joint Chiefs of Staff, Department of Defense Agencies, Joint Service Schools" (Wash., D.C.: DASD (Admin.), July 1978 (May 1979 reprint), which also includes listings of some officials of the DOD. The official directive outlining DOD organization and responsibilities is *Functions of the DOD and Its Major Components* (DOD 5100.1) (Wash., D.C.: DOD), a must for any researcher. DOD directives in the 5100 series also cover the responsibilities in detail of the various Assistant Secretaries of Defense and other higher offices within the Office of the Secretary. A new book available from the GPO is also a valuable reference source and excellent text. *The Department of Defense: Documents on Establishment and Organization 1944-1978* (OSD Historical Office, Wash., D.C.: GPO, 1979) traces the organizational development and history of the DOD through laws, task force reports, directives and correspondence. It contains a number of organizational charts tracing the DOD from its inception to the present day.

Table 13 lists the defense agencies and Unified Commands and gives addresses and telephone numbers of the public affairs offices. The organization and responsibilities of each of the defense agencies is governed by a DOD Directive (listed in parentheses after the name of the agency on Table 13), which should be consulted. Each of the agencies and commands also has its own regulations which outline in more detail its organization. Some agencies print detailed organizational publications.

TABLE 13
DEPARTMENT OF DEFENSE AGENCIES AND UNIFIED COMMANDS*

Defense Advanced Research
 Projects Agency (DOD 5105.47)
1400 Wilson Blvd.
Arlington, VA 22209
(703) 694-3077

Defense Audit Service
 (DOD 5105.48)
1300 Wilson Blvd.
Arlington, VA 22209
(703) 694-9818

Defense Communications
 Agency (DOD 5105.19)
Eighth St. and S. Courthouse Rd.
Arlington, VA 22204
(703) 692-2006

Defense Contract Audit
 Agency (DOD 5105.36)
Building 4, Cameron Station
Alexandria, VA 22204
(703) 274-7319

Defense Intelligence
 Agency (DOD 5105.21)
RTS-2A
Washington, D.C. 20301
(202) 692-5766

Defense Investigative
 Service (DOD 5105.42)
Washington, D.C. 20314
(202) 693-1740

Defense Logistics Agency
 (DOD 5105.22)
Cameron Station
Alexandria, VA 22314
(703) 274-6135

Defense Nuclear Agency
 (DOD 5105.31)
Washington, D.C. 20305
(202) 325-7095

Defense Mapping Agency
 (DOD 5105.40)
U.S. Naval Observatory, Bldg. 56
Washington, D.C. 20305
(202) 254-4140

Defense Security Assistance
 Agency (DOD 5105.38)
Washington, D.C. 20301
(202) 697-0098

CINC, U.S. Atlantic Command
(JO-19) U.S. Naval Base
Norfolk, VA 23511
(804) 444-6294

CINC, U.S. European Command
Public Affairs Officer
APO NY 19128

CINC, U.S. Pacific Command
(JO3/74) PAO
Camp H.M. Smith, HI 96861
(808) 847-1281

CINC, U.S. Readiness Command
Public Affairs Officer
MacDill AFB, FL 33608
(813) 834-4591

CINC, U.S. Southern Command
Public Affairs Officer
APO NY 09826

North American Air
 Defense Command
Public Affairs Officer
Peterson AFB, CO 80914
(303) 635-8911

HQ, United Nations Command/
 United States Forces,
 Korea/EUSA
Public Affairs Officer
APO SF 96301

*DOD Directives on organization and missions of agencies are in parentheses after names of agencies.

The Joint Chiefs of Staff (JCS) receives a lot of attention relating to its functions and responsibilities within the overall military and command system. The official organization and responsibilities of the JCS is outlined in *Organization of the Joint Chiefs of Staff and Relationships with the Office of the Secretary of Defense* (DOD 5158.1) (Wash., D.C.: DOD). A recent study of the history and organizational development of the JCS is *JCS Special Historical Study: A Concise History of the Organization of the Joint Chiefs of Staff: 1942-1978* (Historical Div., Joint Secretariat, Wash., D.C.: DOD, 1979). Two recent books have examined the role of the JCS in the overall national security system: *The Joint Chiefs of Staff: The First Twenty-five Years* (Lawrence J. Korb, Bloomington, IN: Indiana University Press, 1976), and *The Role of the JCS in National Policy* (Wash., D.C.: American Enterprise Institute, August 1978).

Finally, three recent studies have been conducted by the DOD to examine the organizational capability of the Defense Department to conduct its business and satisfy its requirements. These three reports have resulted in various re-examinations of the responsibilities and lines of authority of various components of the DOD. All of these reports are available from the GPO.

Department Headquarters Study: A Report to the Secretary of Defense (Paul R. Ignatius, DOD, Wash., D.C.: GPO, 1978), includes 13 recommendations for management improvements through organizational changes in Department headquarters which would improve efficiency and responsibility.

The National Military Command Structure: Report of a Study Requested by the President and Conducted in the Department of Defense (Richard C. Steadman, DOD, Wash., D.C.: GPO, 1978), examines the role and interaction of the JCS, Unified and Specified Commands, and the National Command Authorities in the time of crisis with recommendations of organizational, management and institutional changes.

Report to the Secretary of Defense of the Defense Agency Review (Theodore Antonelli, DOD, Wash., D.C.: GPO, 1979), a study of the roles, missions, functions and organization of the defense agencies with recommendations for organizational and institutional management changes.

2. The Air Force: The Air Force, like the other armed services, is organized into a headquarters component, numerous operational and support commands, and separate independent agencies reporting to the headquarters. The organizational overview is presented in the *U.S. Government Manual* and the

May issue—the "Air Force Almanac Issue"—of *Air Force Magazine*. The Almanac Issue contains articles on the organization and present status of USAF major commands and separate operating agencies, presents statistics, includes a list and description of bases and R&D centers, and presents charts on the organization of the major commands. The official functions and responsibilities of the Air Force are contained in *Functions of the DOD and its Major Components* (DOD 5100.1). The functions and responsibilities of the headquarters component of the Air Force are contained in *Department of the Air Force, Organization and Functions Chartbook* (HP 21-1). The functions and responsibilities of the major commands and separate operating agencies (see Table 14) are contained in the 23-series (organization and mission) publications (regulations, pamphlets and manuals) of the Air Force. Each major command and separate agency also issues implementing 23-series regulations or pamphlets which prescribes internal organization and functions in more detail. The Air Force regulation that prescribes the organization and mission of each of the major commands is listed in parentheses after the name of the command on Table 14. Two other official publications which outline the organization of the major commands and list the operational units are *Air Force Directory of Unclassified Addresses* (AFM 10-5, Vol. I) and *USAF Command Organization Chartbook* (AFP 23-21), both useful for advanced research. The training establishment and the schools of the Air Force are laid out in *USAF Formal Schools Catalog* (AFM 50-5, Vol. II).

TABLE 14
AIR FORCE MAJOR COMMANDS AND SEPARATE OPERATING AGENCIES*

Air Force Communications
  Command (AFR 23-32)
Scott AFB, IL 62225
(618) 256-4396

Air Force Logistics
  Command (AFR 23-2)
Wright-Patterson AFB, OH 45433
(513) 257-3778

Air Force Systems
  Command (AFR 23-8)
Andrews AFB, MD 20331
(301) 981-4315

Air Training Command
  (AFR 23-6)
Randolph AFB, TX 78148
(512) 652-6307

Alaskan Air Command
  (AFR 23-28)
Elmendorf AFB, AK 99506
(907) 752-2226

Military Airlift Command
  (AFR 23-17)
Scott AFB, IL 62225
(618) 256-5309

Pacific Air Forces
 (AFR 23-27)
Hickam AFB, HI 96853
(808) 449-2834

Strategic Air Command
 (AFR 23-12)
Offutt AFB, NE 68113
(402) 294-5656

Tactical Air Command
 (AFR 23-10)
Langley AFB, VA 23605
(804) 764-5007

United States Air Forces
 in Europe (AFR 23-20)
Ramstein AFB, Ger.
APO NY 09012

USAF Electronics Security
 Command (formerly USAF
 Security Service)
San Antonio, TX 78243
(512) 925-2166

Air Force Accounting and
 Finance Center (AFR 23-26)
Lowry AFB, CO 80230
(303) 320-7741

Air Force Audit Agency
 (AFR 23-38)
Norton AFB, CA 92409
(714) 382-4073

Air Force Intelligence
 Service (AFR 23-45)
Washington, D.C. 20330
(202) 697-3869

Air Force Office of Special
 Investigations (AFR 23-18)
Washington, D.C. 20314
(202) 693-0089

Air Force Inspection and
 Safety Center (AFR 23-15)
Norton AFB, CA 92409
(714) 382-3036

Air Force Test and
 Evaluation Center (AFR 23-36)
Kirtland AFB, NM 87117
(505) 264-5991

Air Force Engineering
 and Services Center
 (AFR 23-4) (AFR 23-35)
Tyndall AFB, FL 32401
(904) 283-2023

Air Force Manpower and
 Personnel Center (AFR 23-33)
Randolph AFB, TX 78148
(512) 652-6141

Air Force Service Information
 and News Center (AFR 23-7)
Kelly AFB, TX 78241
(512) 536-3234

*Air Force Regulations on organization and missions of commands are in parentheses after names of commands.

Two introductory publications provide much information on the background and traditions of the Air Force as well as some useful basic information: *Questions & Answers About Your United States Air Force* (AFP 190-1) and the annual *Air Force Officer's Guide* (MG A.J. Kinney, USAF (ret.), Harrisburg, PA: Stackpole, 1948-). Three bibliographies should also be consulted for background information on the Air Force:

Air Superiority: Selected References
An Aerospace Bibliography
Air Power and Warfare

Finally, there are a number of Air Force manuals which the researcher should be familiar with if he or she is doing work on Air Force doctrine and operations. Table 15 lists the basic documents relating to Air Force doctrine. A complete index appears in *AFR 0-2* (see Section IIIC4).

3. *The Army:* The Army is the largest of the services and also has the most units. Yet less information is available on this

---

TABLE 15
BASIC DOCUMENTS ON
AIR FORCE DOCTRINE

AFM 1-1	USAF Basic Doctrine
AFM 1-3	Doctrine and Procedures for Airspace Control in the Combat Zone
AFM 1-6	Military Space Doctrine
AFM 2-1	Tactical Air Operations: Counter Air, Close Air Support and Air Interdiction
AFM 2-2	Tactical Air Operations in Conjunction with Amphibious Operations
AFM 2-4	Tactical Air Force Operations: Tactical Airlift
AFM 2-5	Tactical Air Operations: Special Air Warfare
AFM 2-6	Tactical Air Operations: Reconnaissance
AFM 2-7	Tactical Air Operations: Tactical Air Control System
AFM 2-11	Strategic Aerospace Operations
AFM 2-12	Tactical Air Operations: Airspace Control in the Combat Area
AFM 2-21	USAF Strategic Airlift
AFM 2-51	U.S. Army/USAF Doctrine for Airborne Operations
AFM 2-53	Doctrine for Amphibious Operations
AFM 3-4	Tactical Air Operations: Tactical Airlift
AFM 3-5	Special Air Warfare Tactics
AFM 3-21	USAF Strategic Airlift

service than on any other of the services. No unclassified listing of Army units is available, although they are available on the other services, and a good organizational manual is not readily available for purchase. An organizational overview of the Army is presented in the *U.S. Government Manual* and the October issue—the "Army Green Book" issue—of *Army Magazine*. The Army Green Book contains articles on the status of the Army commands written by the commanders of those commands, a partial listing of Army units and commands with the names of the commanders, and a gallery of pictures of the top civilian and military officials of the Department. A comprehensive directory of the current weapons of the Army and planned systems in research and development are also included. The official functions and responsibilities of the Army are outlined in *Functions of the DOD and its Major Components* (DOD 5100.1) and *Department of the Army* (AR 10-5). Each of the branches (combat or combat support arms) is outlined in *Branches of the Army* (AR 106). The *Department of the Army Manual* (Office of the Chief of Staff, DA, Wash., D.C.: DA, 1979) (formerly the *DA Greenbook*) is an excellent source of organizational and background information. Although it is not offered for sale, it is available in most Army libraries and at most Army installations.

One source of information on units somewhat unique to the Army is the Table of Organization and Equipment (TO&E) series of publications. These continually-updated publications outline in detail the standard organization and authorization of equipment and personnel in type army units. TO&Es are indexed in *DA Pam 310-3* (see Section IIIC4). Another source of TO&E data is *USA Armor Reference Data, Vol. I: The Army Division, Vol. II: Non-Divisional Organizations* (ST 17-1-1) (U.S. Army Armor School, Dir. of Tng. Dev., Ft. Knox, KY: DA, 1979), an excellent source and presentation of organizational data.

The organization and functions of each of the major commands (see Table 16) of the Army are governed by Army 10-series regulations (the appropriate regulation is cited after the name of the command). Each command also issues its own implementing and amplifying 10-series regulations which discuss in more detail command and subordinate organization and responsibilities. The training establishment and schools of the Army are laid out in *U.S. Army Formal Schools Catalog* (DA Pam 351-4).

Besides the sources listed above, *The Army Officer's Guide* (Lawrence P. Crocker, Harrisburg, PA: Stackpole, 1930-) provides a great deal of information on the background, customs, traditions, responsibilities, organization, missions, personnel

TABLE 16
ARMY MAJOR COMMANDS*

Army Communications
  Command (AR 10-13)
Ft. Huachuca, AZ 85613
(602) 538-2684

Army Criminal Investigation
  Command (AR 10-23)
5611 Columbia Pike, #302
Falls Church, VA 22041
(703) 756-1430

Army Forces Command
  (AR 10-42)
Ft. McPherson, GA 30330
(404) 752-2107

Army Health Services
  Command (AR 10-43)
Ft. Sam Houston, TX 78234
(512) 221-6213

Army Intelligence and
  Security Command
Arlington Hall Station
Arlington, VA 22212
(703) 692-5837

Army Materiel Development
  and Readiness Command
  (AR 10-11)
5001 Eisenhower Ave.
Alexandria, VA 22333
(703) 274-8010

Eighth U.S. Army
Public Affairs Office
APO SF 96301

Army Training and
  Doctrine Command (AR 10-41)
Ft. Monroe, VA 23651
(804) 727-3061

Military District of
  Washington (AR 10-30)
Ft. Lesley J. McNair
Washington, D.C. 20319
(202) 693-1174

Military Traffic Management
  Command (AR 10-18)
Washington, D.C. 20315
(202) 756-2029

USA Europe, and 7th Army
Chief, Public Affairs
APO NY 09403

U.S. Army Japan/IX Corps
Public Affairs Office
APO SF 96343

USA Western Command
Public Affairs Office
Ft. Shafter, HI 96858
(808) 438-9376

Office of the Chief
  of Engineers
Chief of Public Affairs
Pulaski Bldg.
20 Massachusetts Ave., NW
Washington, D.C. 20314
(202) 272-0011

*Army Regulations on organization and missions of commands are in parentheses after names of commands.

procedures, career information, installations, and schools of the Army.

Army operational doctrine is presented in field manuals which discuss operations, the conduct of the battle and operational procedures. The basic doctrinal publications are listed in

Table 17. A complete list and index is *DA Pam 310-3* (see Section IIIC4).

TABLE 17
BASIC DOCUMENTS ON ARMY DOCTRINE

FM 5-100	Engineer Combat Operations
FM 6-20	Fire Support in Combined Arms Operations
FM-6-20-1	Field Artillery Cannon Battalion
FM 6-20-2	Division Artillery, FA Brigade, Field Artillery Section (Corps)
FM 7-7	The Mechanized Infantry Platoon and Squad
FM 7-8	The Infantry Platoon/Squad
FM 7-10	The Light Infantry Rifle Company
FM 7-20	The Infantry Batallion
FM 7-30	The Infantry Brigade
FM 7-85	Ranger Operations
FM 11-50	Combat Communications within the Division
FM 11-92	Corps Signal Communications
FM 17-47	Air Cavalry Combat Brigade
FM 17-95	Cavalry
FM 21-40	NBC Defense
FM 30-5	Combat Intelligence
FM 44-1	U.S. Army Air Defense Artillery Employment
FM 71-1	The Tank and Mechanized Infantry Combat Team
FM 71-2	The Tank and Mechanized Infantry Battalion Task Force
FM 71-3	Armored and Mechanized Brigade Operations
FM 71-100	Armored and Mechanized Division Operations
FM 71-101	Infantry, Airborne and Air Assault Division Operations
FM 90-1	Employment of Army Aviation in a High Threat Environment
FM 90-2	Tactical Deception
FM 90-3	Desert Operations
FM 90-4	Airmobile Operations
FM 90-5	Jungle Operations
FM 90-6	Mountain Operations
FM 90-7	Obstacles
FM 90-10	Military Operations on Urbanized Terrain
FM 90-11	Northern Operations
FM 90-13	River Crossing Operations
FM 90-14	Rear Area Combat Operations
FM 100-5	Operations
FM 100-15	Corps Operations
FM 101-5	Staff Organization and Operation

4. The Navy: A great deal of information is available on Navy organization and forces and on the background and traditions of the Navy. Probably the best overview of any one of the services is *Organization of the U.S. Navy* (NWP-2), a very detailed discussion of the Navy and the operational relationships of its various components. Like the other two major services, the Navy is also detailed in the *U.S. Government Manual* and the May issue—the "Naval Review" issue—of *USNI Proceedings*. *USNI Proceedings* is the best magazine on the U.S. Navy. Besides covering developments in Navy programs, policy, structure and doctrine, *Proceedings* has two annual special reports and the special "Naval Review" issue. The "Naval Review" issue contains articles on the status of the Navy and worldwide naval issues. It also includes a listing of admirals and Marine Corps generals (by rank) with photographs and assignments, and a chronology of naval and maritime events for the preceding year. The January issue contains a special report on the Navy shipbuilding program with a listing of all ships under construction. The January issue also contains a special report detailing changes in the navy force structure from the previous year including commissionings and transfers.

The official functions and responsibilities of the Navy are outlined in *Functions of the DOD and Its Major Components* (DOD 5100.1). The organization and specific functions of the staff offices of Headquarters, U.S. Navy is contained in the *Office of the Chief of Naval Operations (OPNAV) Organizational Manual* (OPNAVINST 5430.48A).

Besides *NWP-2,* the *Standard Naval Distribution List, Part 1, Operating Forces of the Navy, Unified and Specified Commands, U.S. Elements of International Commands* (OPNAV PO9B2-107) (annually issued in July), and *SNDL, Part 2, and Catalog of Naval Shore Activities* (OPNAV P09B2-105) (annually issued in January), are excellent sources of information on Naval force structure, organization and activities. These two publications list Navy and Marine Corps activities and organizations worldwide. The training establishment and schools in the Navy are outlined in *Catalog of Navy Training Courses* (CANTRAC) (NAVEDTRA 10500) and *Bureau of Naval Personnel Formal Schools Catalog* (NAVPERS 91769).

Naval commands issue their own organizational regulations and many organizational instructions outlining subordinate elements of commands. The organizational instructions are in the 5400-series. They are indexed in part in *NAVPUBNOTE 5215* (see Section IIIC4). The applicable instructions appear in parentheses after the name of the command in Table 18.

Because of the fact that Navy tactical organization is dominated by ships, there are additional sources on the Navy that are important for understanding its organization and background. These sources outline specific Naval ships force structure and contain much data and background material.

TABLE 18
NAVY MAJOR COMMANDS AND
OPERATING FORCES

Bureau of Medicine and Surgery (BUMED 5430.4) (OPNAV 5450.178)
Public Affairs Office (Code 0010)
2300 E St., N.W.
Washington, D.C. 20372
(202) 254-4253

Naval Air Systems Command (NAVAIR 5400.1)
Public Affairs Office
1411 Jefferson Davis Highway
Arlington, VA 20360
(703) 692-8373

Naval Education and Training Command
(CNET 5450.4) (OPNAV 5450.194) (CNET 5450.6B)
Public Affairs Office
NAS Pensacola, FL 32508
(904) 452-3613

Naval Electronics Systems Command (NAVELEX 5430.28)
NC #1, 2511 Jefferson Davis Hwy.
Arlington, VA 22202
(703) 692-8954

Naval Facilities Engineering Command
200 Stovall Street
Alexandria, VA 22332
(703) 325-0311

Naval Intelligence Command (NAVINT 5430.2) (OPNAV 5450.181)
4600 Silver Hill Road, Code 221
Washington, D.C. 20389
(202) 763-3552

Naval Materiel Command (NAVMAT 5460.2) (OPNAV 5450.176)
Public Affairs Office (MAT-OOD)
Washington, D.C. 20360
(703) 692-8879

Naval Military Personnel Command/Chief of Naval Personnel
(BUPERS 5400.9)
Public Affairs Office
Washington, D.C. 20370
(703) 694-2815

Naval Sea Systems Command
NC #3, 2531 Jefferson Davis Hwy.
Arlington, VA 20362
(703) 692-1575

Naval Security Group Command
3801 Nebraska Ave., N.W.
Washington, D.C. 20390
(202) 282-0601

Naval Supply Systems Command (NAVSUP 5400.4)
CM #3, 1931 Jefferson Davis Hwy.
Arlington, VA 20376
(703) 695-5351

Naval Telecommunications Command (NAVTEL 5450.37)
  (OPNAV 5450.184)
Administrative Assistant
4401 Massachusetts Avenue, N.W.
Washington, D.C. 20390
(202) 282-0357

Naval Oceanography Command (OPNAV 5450.165)
Hoffman #2, 200 Stovall St.
Alexandria, VA 22332
(703) 325-8778

Operating Forces and Fleets

CINC, U.S. Atlantic Fleet
Norfolk, VA 23511
(804) 444-6294

Military Sealift Command
4228 Wisconsin Ave., N.W.
Washington, D.C. 20390
(202) 282-2808

CINC, U.S. Pacific Fleet
Pearl Harbor, HI 96860
(808) 422-9922

Cdr., 7th Fleet
FPO SF 96601

U.S. Naval Forces, Europe
Public Affairs Office (Box 13)
FPO NY 09510

Cdr., 2nd Fleet
FPO NY 09501
(804) 444-2422

Cdr., 3rd Fleet
Pearl Harbor, HI 96860
(808) 472-8371

Cdr., 6th Fleet
FPO NY 09501

Naval Instructions on organization and missions of commands are in parentheses after names of commands.

Ships and Aircraft of the U.S. Fleet (Norman Polmar, ed., Annapolis, MD: USNI, 1981), is an excellent source of information, presenting in detail the characteristics and identification of the ships and aircraft used by the U.S. Navy. Another excellent source is an unpublished quarterly pamphlet issued by the Navy Program Information Center, "Listing of Active Fleet, Naval Reserve Force and MSC-Naval Fleet Auxiliary Force Ships by Fleet," available through the Navy PAO. One should also use the three special issues of *USNI Proceedings* for information on Naval forces and ships.

Besides the sources mentioned above, there are a number of good sources on naval traditions, operations, personnel policy, procedures and assignments. One introductory book about the Navy worth obtaining is *Naval Orientation* (NAVPERS 10900-83), a comprehensive look at many aspects of the Navy with good introductory and explanatory material. The USNI publishes three books that also contain much useful background information:

Naval Officer's Guide (Arthur A. Ageton and William P. Mack, Annapolis, MD: USNI, 1943-), a periodically updated introduction to the Navy directed at Naval officers.

Division Officer's Guide (John V. Noel, et al., Annapolis, MD: USNI: 1976), a manual of shipboard operations, supervision and responsibilities.

The Blue Jacket's Manual (Bill Wedertz, Annapolis, MD: USNI, 1978), an enlisted men's guide and introduction to the Navy.

The official regulation which prescribes uniform organizational structure and responsibilities, regulations, and readiness requirements of U.S. Navy ships and combat organizations is *Standard Organization and Regulations of the U.S. Navy* (OPNAVINST 3120.32A).

There are also a few sources on Naval aviation. *Ships and Aircraft of the U.S. Fleet* is the best source on equipment, while a number of the general sources listed above also contain sections relating to Naval aviation. The USNI publishes another guide, *The Naval Aviation Guide* (VADM Malcolm W. Cagle, USN (ret.), Annapolis, MD: USNI, 1976), which provides information on the background, doctrine, techniques and procedures of Naval aviation.

Finally, as with the other services, there are the official doctrinal manuals which are of great use for advanced research. What is somewhat unusual, though, is that although more material is available on Navy organization and forces than on the other services, most of the Naval doctrinal manuals, specifically those which detail operations at sea, are classified. The major unclassified manuals are listed in Table 19.

TABLE 19
BASIC DOCUMENTS ON NAVAL DOCTRINE

NWP-1	Strategic Concepts of the U.S. Navy
NWP-2	Organization of the U.S. Navy
NWP-3	Naval Terminology
NWP-10-2	Law of Naval Warfare
NWP-13	Doctrine for Navy/Marine Corps Joint Riverine Operations
NWP-14	Replenishment at Sea
NWP-22	Doctrine for Amphibious Operations
NWP-22-1	The Amphibious Task Force Plan
NWP-22-2	Supporting Arms in Amphibious Operations
NWP-22-5	The Naval Beach Group
NWIP-22-6	Amphibious Embarkation
NWP-50	Air Organization for the Support of Naval Doctrine
AMP-7	Helicopter Mine Countermeasures Manual
ATP-27	Offensive Air Support Operations

5. *The Marine Corps:* The Marine Corps is not technically an armed service, but since it has a separate management headquarters which handles public affairs and publications, it is separated from the Navy in this guide. Much of the material which presents an organizational overview of the Navy also includes sections on the Marine Corps. Aside from those sources, basically four sources present more detailed and authoritative information on staff and force organization and responsibilities. The *HQMC Organizational Manual* (HQ OP5400.18) presents the organization and responsibilities of the various offices of Headquarters, Marine Corps. The *Marine Corps Manual* (Wash., D.C.: USMC) also discusses organization but includes much data on the mission, background and responsibilities of the Marine Corps. *FMF Organization* (ECP 1-4) outlines tactical Marine unit organization and the *List of Marine Corps Activities* (MCO P-5400.6) is an annual directory of Marine Corps activities and units.

Two additional sources are helpful for background

information. *Creating a Legend: The Complete Record of Writing about the United States Marine Corps* (John B. Moran, Chicago: Moran Andrews, 1973) is an exhaustive bibliography, and *The Marine Officer's Guide* (COL Robert D. Heinl, Jr., USMC (ret.), Annapolis, MD: USNI, 1977) outlines Marine Corps organization and functions as well as tradition, protocol and personnel procedures.

There are basically three major commands in the Marine Corps which may be sources of additional information. These three commands supervise the training establishment and the operational units of the Marine Corps.

HQ, Fleet Marine Force, Atlantic
Norfolk, VA 23511

HQ, Fleet Marine Force, Pacific
Camp H.M. Smith
Honolulu, HI 96861

Marine Corps Development and Education Command
MCB Quantico, VA 22134

Doctrine for the Marine Corps is outlined in *Fleet Marine Force Manuals* (FMFMs) and *Landing Force Manuals* (LFMs). Table 20 lists these manuals. They are an important source of operational information.

6. The Reserves: There are seven reserve components (including the Coast Guard Reserve, not covered here) of the U.S. military, all of which have separate structures, management headquarters and information sources. There is not an overall guide to the Reserve Forces, but many of the general sources on the armed services have sections on their reserve components. The *Reserve Forces Almanac* (Wash., D.C.: Uniformed Services Almanac, Inc., 1974-) is an annual commercial publication that primarily reports personnel pay, benefits and policies but also contains some useful background information on organization and history.

The National Guard Bureau, a joint Army-Air Force management headquarters for the Army and Air National Guard components, has a separate public affairs office and publishes regulatory material. The organization and status of the National Guard is outlined in the *Annual Review, FY__, Chief, National Guard Bureau* (Wash., D.C.: NGB). The official functions of the National Guard Bureau are outlined in *Organization and Functions of the National Guard Bureau* (AR 130-5/AFR 45-17). The *National Guard Almanac* (Wash., D.C.:

TABLE 20
BASIC DOCUMENTS ON MARINE CORPS DOCTRINE

LFM 01	Doctrine for Amphibious Operations
LFM 02	Doctrine for Landing Forces
FMFM 0-1	MAGTF Doctrine
FMFM 1-38	Sniping
FMFM 2-1	Intelligence
FMFM 2-2	Amphibious Reconnaissance
FMFM 3-1	Command and Staff Action
FMFM 3-3	Helicopterborne Operations
FMFM 4-1	Combat Service Support for Marine Air-Ground Task Forces
FMFM 4-2	Amphibious Embarkation
FMFM 4-3	Landing Support Operations
FMFM 4-4	Engineer Operations
FMFM 5-1	Marine Aviation
FMFM 5-3	Assault Support
FMFM 5-4	Offensive Air Support
FMFM 5-5	Anti-Air Support
FMFM 6-1	Marine Division
FMFM 6-2	Marine Infantry Regiment
FMFM 6-3	Marine Infantry Battalion
FMFM 6-4	Marine Rifle Company/Platoon
FMFM 6-5	Marine Rifle Squad
FMFM 7-2	Naval Gunfire Support
FMFM 7-4	Field Artillery Support
FMFM 8-1	Special Operations
FMFM 8-4	Doctrine for Navy/Marine Corps Joint Riverine Operations

Uniformed Services Almanac, Inc., 1975-) is an annual review similar to the *Reserve Forces Almanac* which contains information on pay, benefits, history, background, organization and personnel. Other sources of information on the structure of the individual State National Guards (the National Guard becomes a fully federally controlled armed force after mobilization) are the annual and biennial reports of each state's Adjutant Generals (top National Guard officer in each state). These report on the organization and operations of the National Guard units in the state system. Information on the State National Guard is often contained in state "Blue Books" or "Legislative Manuals" which describe the organization of the state government. Finally, the National Guard Bureau Public Affairs Office issues a series of fact sheets on National Guard organization, background and policies.

>Chief, National Guard Bureau
>Public Affairs Office
>Washington, D.C. 20310
>(202) 695-0421

The Army Reserve also has a separate PAO and information system. Some information on the Army Reserve is contained in the *Reserve Forces Almanac. A Guide to the Reserve Components of the Army* (DA Pam 135-3) contains a good explanation of the organization and functions of the Army Reserve and the Army National Guard, and *Army Reserve: Mission, Organization, Training* (AR 140-1) is the official regulation on the mission of the Army Reserve. Another source with much information on the Army Reserve components is *The Role of the Reserve in the Total Army: A Bibliographic Survey of the U.S. Army Reserve* (Dept. of the Army, Army Library, Wash., D.C.: GPO, 1977), an annotated and selected bibliography dealing with the Army Reserve.

>Office of the Chief of Army Reserve
>Public Affairs Office
>Washington, D.C. 20310
>(202) 695-1527

Organizational information on the Naval Reserve is contained in most of the sources mentioned under the Navy, specifically those sources which list Naval units. The official functions and responsibilities of the Naval Reserve are contained in *Function of the Chief of Naval Reserve* (OPNAV 5450.186A).

>Chief of Naval Reserve
>4400 Dauphine St. (Code 004)
>New Orleans, LA 70146
>(504) 948-1240

Information on the Air Force Reserve is also contained in a number of the sources mentioned under the Air Force. The official regulation on the organization and mission is *HQ, Air Force Reserve* (AFR 23-1).

Air Force Reserve
Public Affairs Office
Robins AFB, GA 31098
(912) 926-5202

The best source of information on the Marine Corps Reserve is the 15-series *Education Center Publications* (ECPs) which deals with various aspects of the Marine Corps Reserves. The sources of additional information available from the public affairs offices of the Marine Corps Reserve are split into two major reserve commands, each commanding the ground or air units.

Commander, 4th Mar. Div., USMCR
4400 Dauphine St.
New Orleans, LA 70146

Commander, 4th MAW/MARTC
4400 Dauphine St.
New Orleans, LA 70146

B. DEFENSE POLICY AND POSTURE

The background sources of information on the organization and functions of the U.S. military presented in the previous section are intended to familiarize the researcher with the institutions and legacy of the Department of Defense and the armed services. U.S. military policy, which determines the structure and operations of the military, and the actual posture and readiness of the military forces, which in turn influence the ability to carry out military objectives, are discussed in this section. Some basic sources on U.S. national security and foreign policy are discussed, and sources which report the current policy, posture and readiness of the armed forces are presented.

1. U.S. National Security Policy: In this section, sources on both the background of post-World War II national security policy and the current overall defense and foreign policy are discussed. U.S. national security policy, as presented annually by the President, the Secretaries of Defense and State, and the ranking military officers, is well documented. In addition, the conduct of that policy—including any deviance or modification to react to crises or new circumstances or to satisfy the needs of special interests—is well covered in the media, the specialized periodicals, and Congressional hearings and studies. A researcher in this field has a tremendous amount of material available to analyze. This section highlights some of the basic and most useful sources so that one may both understand and monitor U.S. overall policy and the ability to take action in support of it.

(a) Background: A number of good sources provide an overview of U.S. national security policy. Much has been written on the subject and two bibliographies: *National Security Affairs: A Guide to Information Sources* and *American Defense Policy Since 1945: A Preliminary Bibliography*, are good tools to trace that writing. Two additional books provide much lasting data and background material. *American Defense Policy* (John E. Endicott and Roy W. Stafford, Jr., eds., Baltimore, MD: Johns Hopkins Univ. Press, 1977), a reader in its fourth edition, provides a documentary anthology of U.S. military policy and programs and the evolution of defense strategy, policy and institutions. *Force Without War: U.S. Armed Forces as a Political Instrument* (Barry M. Blechman and Stephen S. Kaplan, Wash., D.C.: Brookings, 1978) is a comprehensive study of the use of American military force (215 instances are examined since World War II) to influence events, and an examination of the actual effectiveness of the use of military power. It is a valuable reference work.

A number of recent books may also provide a background and perspective for understanding U.S. national security policy. Two readable studies provide information and data on current policy and on the politics of U.S. military policy and forces. *U.S. Defense Policy: Weapons, Strategy and Commitments* (Wash., D.C.: Congressional Quarterly, 1978) and *U.S. Defense Policy* (2d Ed.) (Wash., D.C.: CQ, 1980) discuss defense issues before Congress in the 1973-1977 and 1977-1980 periods and review legislation and action on that policy. They are non-biased and include selected documents. *The Price of Defense: A New Strategy for Military Spending* (Boston Study Group, New York: New York Times Books, 1979) is another useful introduction to the range of current national security issues, questioning the size of U.S. military forces, certain military policies, and military expenditures beyond the need of forces necessary to carry out U.S. national security policy.

Two anthologies of articles criticize the size of U.S. defense programs and U.S. policy for not being realistic in light of the perceived world situation and the size of Soviet military forces and influence. They examine policies relying more heavily on military forces and power from various perspectives, and are good overviews presenting the prevailing conservative view of U.S. military policy and power: *The Changing World of the American Military* (Franklin D. Margiotta, ed., Boulder, CO: Westview, 1979) and *U.S. National Security Policy in the Decade Ahead* (James E. Dornan, ed., New York: Crane, Russak, 1978).

Other works on the strategic military balance and specific aspects of U.S. military forces are discussed in subsequent sections.

(b) Current Defense Posture: The current posture of the U.S. military is presented in a number of annual Defense Department publications outlining policy, programs and spending and is examined in Congressional hearings (see Table 5) which authorize and approve military spending. Periodicals and newsletters are also a primary and excellent source on current defense policy.

The overall programs of the Defense Department, the status of the military, the Defense Department assessment of the world military situation and the five year projection in military spending, programs and procurement is presented in two basic documents with which every researcher should be familiar: *United States Military Posture for Fiscal Year 19__* (aka "The Joint Chief's of Staff Military Posture Statement" (OJCS, Wash., D.C.: GPO), an annual document describing U.S. forces, regional balances, the Soviet military and the U.S.-Soviet balance; and *Department of Defense Annual Report* (aka "The Secretary's Posture Statement") (DOD, Wash., D.C.: GPO, 1963-), containing analysis of the world situation, the posture of the U.S. military and the major military powers and presenting the fiscal year defense budget and programs in light of the world analysis. It also includes a five year projection of programs, systems and budgets.

These two basic documents are supplemented by about fifteen more specific Defense Department reports (on personnel, research and development, procurement, military construction, etc.) which form the basis of testimony before Congress and describe in detail military programs. These publications are discussed in subsequent sections.

The hearings themselves are the best source on the current military posture, the status of strategic and general purpose forces and the strategic, theater and regional balances. Section IIIB discusses the hearings and subsequent sections refer back to specific portions of these documents.

Four annual assessments by non-DOD groups analyze the defense budget and defense programs and spending. *Setting National Priorities: The 19__ Budget* (Joseph A. Pechman, ed., Wash., D.C.: Brookings, 1970-) is an annual analysis of the overall federal budget and includes a section on the defense budget. *Arms, Men and Military Budgets: Issues for Fiscal Year 19__* (Francis R. Hoeber and William Schneider, Jr., eds., New York: Crane, Russak, 1976?-) is an annual conservative assessment. A third assessment, also conservative, is "The FY 19__-19__ Defense Program: Issues and Trends" (Lawrence J. Korb, in *AEI Foreign Policy and Defense Review*, 1977-), an annual issue of the AEI quarterly journal. Finally, a fourth authoritative and non-biased analysis of the defense budget is an annual issue brief prepared by the Congressional Research Service, *Defense Budget: FY__* (Wash., D.C.: CRS).

Periodicals and newsletters are essential sources for following defense policy, programs and issues. While some periodicals cover overall defense issues, others cover issues of a specific branch of service. (These are discussed in later sections dealing with the current posture of the armed services.) Many of these periodicals, besides reporting military news and current defense activities and programs, also contain annual reports and assessments of the defense budget. The seven major periodicals that report defense news and contain information on programs, research and development, procurement, resource management, costs and policies are *Air Force Magazine, Armed Forces Journal International, Army, Aviation Week and Space Technology, Marine Corps Gazette* and *USNI Proceedings.* Other periodicals covering defense-wide issues are *Defense Electronics, Defense Transportation Journal, Military Electronics/Countermeasures, The Military Engineer, Military Market Magazine, National Defense, The Review,* and *Signal.* Table 2 lists magazines covering worldwide military affairs, a major part of the reporting covering U.S. programs and policy. Finally, there are a number of daily newsletters which report U.S. defense programs, contracting, procurement, R&D and spending. These newsletters are services aimed mostly at corporate customers and concentrate on news of interest to defense industry, contractors and lobbyists. Table 21 contains a list of the newsletters dealing with the U.S. military and U.S. military programs and policies.

(c) U.S. Foreign Policy: U.S. foreign policy governing the general relations of the U.S. with other countries, participation

TABLE 21
NEWSLETTERS ON CURRENT DEFENSE POSTURE AND PROGRAMS

Aerospace Daily
Aerospace Intelligence
Defense & Foreign Affairs Daily
Defense and Economy World Report and Survey
Defense Daily
Defense Survey
Defense Week
The Government Contractors Communique
International Defense Business
Military Research Letter
Missile/Ordnance Letter
Renegotiation/Management Letter
Underwater Letter

---

in bilateral and multilateral international organizations, aid, and arms control and disarmament is administered by the State Department. This is a central area to understanding military and strategic affairs and there are a number of general sources of importance to researchers in addition to some specific sources mentioned later under other sections. The basic documents on foreign policy are the annual reports of the State Department, the Arms Control and Disarmament Agency and the Agency for International Development. The annual report of the Secretary of State, *U.S. Foreign Policy: A Report of the Secretary of State* (Wash.: D.C. GPO), is the official report and along with the hearings on U.S. Foreign Relations and Aid (see Table 5), presents a tremendous resource on U.S. relations with foreign countries. The overlap between foreign policy and military policy, specifically when referring to external policy, is great and many of the sources on military assistance and general military and strategic affairs cover both areas.

A valuable reference book on U.S. foreign policy, *Legislation on Foreign Relations through 19__, Vol. I-II, Current Legislation and Related Executive Orders, Vol. III: Treaties and Related Materials* (Report prepared by the CRS as House Foreign Affairs Committee/Senate Foreign Relations Committee Print, Wash., D.C.: GPO), is a compilation of foreign relations legislation, treaties and laws regulating the conduct of foreign affairs in effect as of January 1 of the year of publication.

Another good source which reviews the conduct of U.S. foreign policy annually is the *United States in World Affairs, 19__* (Council on Foreign Relations, New York: Simon and Schuster, 1968-), a collection of essays and a detailed chronology.

Current issues in U.S. foreign policy are covered in International Relations Journals (see Table 1). The *Department of State Bulletin,* the official magazine of the State Department, is a source of press releases, statements, addresses and articles on international affairs and foreign relations. It should not be overlooked as a useful source of information, and it is widely accessible.

(d) The U.S.-Soviet Military Balance: Probably no other area dealing with military and strategic affairs receives more attention than the U.S.-Soviet military balance. The IISS *Strategy Survey* and the *SIPRI Yearbook* are probably the closest things to authoritative and unbiased assessments of the strategic and theater balances. The strategic journals listed in Table 2, are also good sources but most are biased one way or the other on the state of the balance.

The researcher may desire sources of data and information which present good material for making independent assessments of the state of the U.S.-Soviet military balance. A short CRS study, *U.S./Soviet Military Balance: A Frame of Reference for Congress* (A study by the CRS for the Comm. on Armed Services, U.S. Senate, Wash., D.C.: GPO, 1976), is an analysis of the U.S.-Soviet military balance and the characteristics and actors of the Soviet and U.S. militaries affecting the balance.

Three additional studies are a source of a great deal of data and commentary on the balance. They provide the basis for almost every quantifiable aspect of U.S. and Soviet military forces and much background information. These studies, by a senior defense analyst at the Congressional Research Service present with hard data hawkish views of the military balance:

U.S.-Soviet Military Balance: Concepts and Capabilities, 1960-1980 (John M. Collins, Wash., D.C.: McGraw-Hill Publishing, 1980).

American and Soviet Military Trends since the Cuban Missile Crisis (John M. Collins, Wash., D.C.: Georgetown University (CSIS), 1978).

Imbalance of Power (John M. Collins and Anthony Cordesman, San Rafael, CA: Presidio Press, 1978).

Another study, *The Soviet Military Buildup and U.S. Defense*

Spending (Barry M. Blechman, et. al., Wash., D.C.: Brookings, 1977), examines sustained improvements in the Soviet military, the conventional military balance in regions of the world and U.S. programs needed to counter the Soviet build-up.

The annual hearings on the defense budget examine various aspects of the U.S.-Soviet military balance. The *JCS Posture Statement*, the *DOD Annual Report*, and the *Annual DOD RDA Report* include comprehensive assessments of the balance. Congressional hearings have also been held regularly by the Joint Economic Committee to examine the state of the Soviet economy and the allocation of Soviet resources to military programs. These hearings, *Allocation of Resources in the Soviet Union and China*, are another valuable resource.

Finally, a number of Congressional Budget Office studies that examine U.S. strategic forces are good sources of information on the military balance:

Planning U.S. Strategic Nuclear Forces for the 1980s (June 1978)
Retaliatory Issues for the US. Strategic Nuclear Forces (June 1978)
Counterforce Issues for the U.S. Strategic Nuclear Forces (Jan. 1978)
U.S. Strategic Nuclear Forces: Deterrence Policies and Procurement Issues (Apr. 1977)
SALT and the U.S. Strategic Forces Budget (June 1976)

2. Air Force Current Posture: The current posture of the Air Force is outlined annually before Congress as part of the hearings on the defense budget. The portion of the hearings normally dealing with the actual organizational and material readiness and posture of the armed services are the "operations and maintenance hearings" of the individual hearings on the posture and budgets of the armed services (see Section IVC on the Defense Budget for more information). One should not overlook the annual hearings as a source for in-depth information on the posture and readiness of any one of the armed services.

Background information on current issues dealing with the forces and posture of the Air Force can be obtained from a number of excellent studies on specific aspects of the Air Force. Three studies have been conducted recently on the U.S. tactical air forces: *U.S. Tactical Air Power: Mission, Forces and Costs* (William D. White, Wash., D.C.: Brookings, 1974), *Planning U.S. General Purpose Forces: The Tactical Air Forces (*CBO, Wash.,

D.C.: GPO, Jan. 1977), and *U.S. Tactical Air Forces: Overview and Alternative Forces, Fiscal Years 1976-1981* (CBO, Wash., D.C.: GPO, Apr. 1976). These studies review the missions and status of the tactical air forces and provide various alternatives for future procurement and strategies.

Airlift capabilities of the United States have been the subject of two recent studies, both of which provide much background data for a topic that is of great interest today. *U.S. Airlift Forces: Enhancement Alternatives for NATO and non-NATO Contingencies* (CBO, Wash., D.C.: GPO, Apr 1979) outlines current Air Force airlift programs and their rationale. Hearings were also held before the House Armed Services Committee in September 1977, providing much background information for the researcher on the capability of U.S. airlift forces: *Posture of U.S. Military Airlift* (Hearings before the Comm. on Armed Services, U.S. House, Wash., D.C.: GPO, 1977).

Another source of information is *Modernizing the Strategic Bomber Forces: Why and How* (Alton H. Quanbeck and Archie L. Wood, Wash., D.C.: Brookings, 1976), a study originally done in light of an upcoming decision on the abandoned B-1. It is still valuable because of the continuing questions about the future of U.S. bomber and strategic forces.

Besides the hearings and some timely and cognizant studies, the best source of continuing information on the posture of the Air Force are periodicals and newsletters. In Table 7, the periodicals published by the Air Force are listed. Some of these periodicals are good sources on the current posture and programs of the Air Force. *Air Force Policy Letter for Commanders*, the *Supplement* and *Airman* are general reviews of what is going on in the Air Force. *Air University Review* is a scholarly journal dealing with the broad issues of aerospace, air strategy and operation of the present Air Force. But official publications are by no means the best sources of information on the Air Force. *Air Force Magazine* and *Aviation Week and Space Technology* are by far the most often read and quoted magazines dealing with the current posture of the Air Force. They include many special issues which make them especially valuable reference sources. Other periodicals and newsletters worth special mention are *Air Force Times,* which concentrates on news of interest to Air Force personnel such as personnel policy and benefits, but also includes some information on exercises and organization; and the newsletters dealing with aviation and aerospace listed in Table 21. Much information on the U.S. Air Force is also included in many of the periodicals which cover worldwide military affairs (see Table 2). Actually, there is more written on the Air Force, specifically aircraft and missiles, than on any other area in the military magazines.

3. Army Current Posture: The current posture of the Army is outlined in two official documents published by the Army: *The Posture of the Army and Department of the Army Budget Estimates for Fiscal Year 19__* (Wash., D.C.: DA) and *Equipping the United States Army: A Statement to Congress on the FY__ Army RDTE and Procurement Operations* (Wash., D.C.: DA). These two publications form the backbone of the Army presentation to the Congress as part of the annual hearings on the defense budget and the "operations and maintenance" and posture of the Army. The hearings themselves contain much information on the organization and posture of the Army, the programs and the current and future status of Army systems and weapons.

Two CBO studies provide an understanding of some present issues facing the Army: *Planning U.S. General Purpose Forces: Army Procurement Issues* (Dec. 1976) and *U.S. Army Force Design: Alternatives for Fiscal Years 1977-81* (July 1976).

Periodicals should also be consulted for current information on Army programs and posture. Table 8 lists the periodicals published by the Army. A number of these magazines are directed at the different combat arms and are produced by the central schools. These magazines contain articles about operational organization, training, doctrine, tactics, and equipment and news items about personnel matters, Army-wide developments and R&D. Those doing advanced research on Army operational doctrine and current posture should consult these magazines:

Air Defense Magazine
Armor
The Army Communicator
Army Logistician
The Engineer
Field Artillery Journal
Infantry
Military Intelligence
United States Army Aviation Digest

The general news magazines published by the Army are *Commanders Call, EurArmy Magazine,* and *Soldiers.* Two scholarly journals, *Military Review* and *Parameters,* are also published by the Army with longer articles examining issues of doctrine, operations, programs and overall Army and defense policy. As with the other services, good commercial publications provide much information on the current Army posture. The three major publications are *Army, Army Aviation* and *Army Times*. *Army* and *Army Times* cover general military affairs and personnel policy and programs and are important sources for the advanced researcher. The various newsletters (see Table 21) and the magazines dealing with worldwide affairs (see Table 2) also serve as sources on current Army posture.

4. *Navy Current Posture:* There are many sources of current Navy posture and a number of recent studies on Navy force structure and the future of the Navy. The annual *CNO Report* (A Report by the Chief of Naval Operations concerning the Fiscal Year 19__ military posture and budget of the United States Navy) (Wash., D.C.: DN) summarizes Naval programs and posture and is one of the basic documents in support of the Navy portion of the defense budget. The hearings on the defense budget provide much more information on the Navy, and since the future of the Navy and Navy force structure has been hotly debated in recent years, they are particularly good sources. Other hearings have also been held recently dealing with the Navy, including some on Navy shipbuilding programs and seapower.

The debate over the future of the Navy has resulted in a number of CBO studies:

Navy Budget Issues for Fiscal Year 1980 (Mar. 1979)
U.S. Naval Forces: The Peacetime Presence Mission (Dec. 1978)
U.S. Projection Forces: Requirements, Scenarios and Options (Apr. 1978)
The U.S. Sea Control Mission: Forces, Capabilities and Requirements (June 1977)
Planning U.S. General Purpose Forces: The Navy (Dec. 1976)
U.S. Naval Force Alternatives (Mar. 1976)

The Navy response to one of these studies was printed as a House Armed Services Committee Print, *U.S. Navy Analysis of CBO Budget Issue Paper "General Purpose Forces: Navy"* (Report of the House Armed Services Committee, 95-1, Wash., D.C.: GPO, 1977). The CRS has also prepared some Issue Briefs on Navy issues, including *Navy Aircraft Carriers, Trident Program* and *Navy Shipbuilding.*

Two recent studies may also provide some useful background information: *White Paper on Defense: A Modern Military Strategy for the United States* (1978 ed.) (Robert Taft, with Senator Gary Hart, prepared with the assistance of William S. Lind, Wash., D.C.: processed, May 15, 1978) and *The Future of U.S. Naval Power* (James A. Nathan and James K. Oliver, Bloomington, IN: Indiana University Press, 1979). These studies examine U.S. conventional seapower and the forces and doctrine needed for performing various seapower strategies.

Periodicals are excellent sources of information on the current posture of the Navy. Table 10 lists periodicals published by the Navy. Two of those publications, *Naval Aviation News* and *Surface Warfare* are excellent sources of background information on Naval organization, activities and programs. These two periodicals, similar to the magazines mentioned under the Army, contain organizational profiles and news items directed at Naval personnel in the aviation and surface warfare fields. Another periodical, *All Hands,* reports general news on the Navy. *Naval War College Review,* the scholarly journal of the Navy, contains articles on military sciences, naval doctrine, policy and operations.

In addition to the periodicals published by the Navy, a number of commercial publications deal with the Navy and naval affairs. *USNI Proceedings* has already been discussed and is the best magazine dealing with the U.S. Navy. *Navy Times* not only covers personnel policy and benefits but reports exercises, organizational changes and other news items useful for advanced research. *Seapower,* the publication of the Navy League of the United States, is another Navy-related publication which advocates a stronger Navy and seapower strategy. In addition to the magazines dealing specifically with the U.S. Navy, many of the newsletters (see Table 21) report naval issues and many of the worldwide military magazines (see Table 2) contain frequent articles on the U.S. Navy, Naval programs and policy, ships and weapons.

5. Marine Corps Current Posture: The Congressional hearings on the defense budget are the best source of information on the current posture of the Marine Corps. A number of the studies dealing with the Navy referenced above also deal with the Marine Corps, specifically those dealing with procurement (which is done by the Navy) and amphibious lift (which is operated by the Navy).

The future of the Marine Corps has been debated in recent years and many studies provide some background material for this debate: *Structuring the Marine Corps for the 1980s and 1990s* (John Grinalds, Wash., D.C.: NDU (National Security Affairs Monograph Series 78-6) GPO, 1978), *Where Does the Marine Corps Go From Here?* (Martin Binkin and Jeffrey Record, Wash., D.C.: Brookings, 1976), *Marine Amphibious Forces: A Look at Their Readiness, Role, and Mission* (Wash., D.C.: GAO, Feb. 5, 1979), and *The Marine Corps in the 1980s: Prestocking Proposals, the RDF and Other Issues* (Wash., D.C.: CBO, 1980). Many articles in periodicals have also dealt with the future of the Marine Corps and should be consulted.

The Marine Corps does not publish magazines like the other services but relies instead on two magazines of the Marine Corps Association, *Leatherneck* and *Marine Corps Gazette*. *Leatherneck* is directed at enlisted Marines and includes personnel and general Marine Corps news items. *Marine Corps Gazette* is directed at Marine Corps officers and is intended to be the scholarly journal of the Marine Corps covering organization, activities and doctrine and tactics. Many of the periodicals dealing with the Navy also include material dealing with the Marine Corps and many of the magazines reporting worldwide military affairs (see Table 2) also contain occasional articles on the Marines.

6. Reserve Forces Current Posture: Although some information on the current posture of the reserve components is separated from the parent component, most of the sources dealing with the Army, Navy, Air Force and Marine Corps also deal with their reserve components. The Congressional hearings on the defense budget contain sections dealing with the reserves and should be consulted. The Reserve Force Policy Board submits an annual report, *Annual Report of the Reserve Forces Policy Board, Fiscal Year__* (Wash., D.C.: DOD ASD (MRAL)), which reviews the work of the board and the state of the reserves. The *Annual Report of the Secretary of Defense on Reserve Forces* was discontinued in 1979 and the information was incorporated into the *Department of Defense Annual Report* and the *Manpower Requirements Report,* although not in as much detail.

Much information is available on the National Guard, in the state annual or biennial reports and the *Annual Review, Chief, National Guard Bureau.* Most of the other components also have some sort of unpublished annual review which is normally incorporated into the annual hearings.

Two studies on the reserves might also be helpful for background information: *Improving the Readiness of the Army*

Reserve and National Guard: A Framework for Debate (CBO, Wash., D.C.: GPO, Feb. 1978) and *U.S. Reserve Forces: The Problem of the Weekend Warrior* (Martin Binkin, Wash., D.C.: Brookings, 1974).

Finally there are the periodicals which deal specifically with the reserve components. Each one of the components has a government or commercial publication which reports news, personnel policy and other policies dealing with the reserves. They are *Air Reservist, Army Reserve Magazine, Continental Marine and Digest, National Guard, Naval Affairs* and *The Officer*.

C. THE DEFENSE BUDGET

The defense budget is the central element of military policy, future plans and Congressional oversight of the executive branch and the military. The features of the defense budget—after a period of formulation, review, presentation, oversight and approval—control the actual current size and composition of U.S. military forces and the short- and long-term future composition and focus of military forces and programs. The defense budget goes through a complicated system of formulation and approval within the executive branch and then is examined, perhaps altered, and approved by the legislative branch. Understanding the defense budget, therefore, is essential to doing advanced research on military and strategic affairs.

In this section, the works which directly document and support the defense budget are presented. Many of these documents are very technical and unnecessary for most research. The budget process, including formulation of the initial budget within the Defense Department and the annual calendar and cycle of oversight and review is also presented. An understanding of these elements of the budgets is necessary both to follow the status of the defense budget and to analyze the politics of military expenditure and strategy.

1. The Budget Process and Budget Formulation: A number of works explain the overall federal budget process and the specific defense budget process. Two general studies describe the workings of the budget system: *Congress and the Budget* (Joel Havemann, Bloomington, IN: Indiana Univ. Press, 1978) and *Governing Budgeting: Theory, Process, Politics* (Albert C. Hyde and Jay M. Shafritz, Oak Park, IL: Moore, 1978), but these are by no means the extent of the works on this subject. A few good studies exist on defense budgeting which can also serve as valuable background for research. *Defense Politics: A Budgetary Perspective* (Arnold Kantor, Chicago: Univ. of Chicago Press, 1979) is the latest of a number of works by an expert on defense budget politics, which analyzes the bargaining process in defense budgets and president/bureaucracy relations during the Eisenhower through Johnson years. *Defense Budgeting: The British and American Cases* (Richard Burt, London: IISS (Adelphi Paper 112), 1975) and *Real Growth and Decline in Defense Operating Costs: Fiscal Year 1978* (CBO, Wash., D.C.: June 1977) are good case studies on specific aspects of defense budgeting. The Brookings annual studies, *Setting National Priorities,* also provide good background information on defense budgeting and the overall process.

A number of Defense Department publications provide a great deal of data on budget formulation and on the planning, programming and budgeting system (PPBS) of the military. One very readable explanation of the budget process, PPBS, budget formulation, enactment and execution is *The Air Force Budget* (AFP 172-4), an annual Air Force publication. Another publication which explains the PPBS system and the internal Defense planning, programming and budgeting procedures in detail is the *Planning, Programming and Budgeting System Handbook* (Wash., D.C.: DA, Office of the Chief of Staff, 1979). The *PPBS Handbook* is highly recommended for advanced research.

Other documents of the Defense Department like the *DOD Budget Guidance Manual* (DOD 7110-1-M) and the Budget or Comptrollers Manuals of the services (e.g., *USAF Budget Manual* (AFM 172-1), *Army Comptrollers Handbook* (DA Pam 37-4), *Navy Comptrollers Manual,* et. al.) are valuable resources for understanding the intricacies of the budget process. The cost planning handbooks (e.g., *Army Force Planning Cost Handbook* (Wash., D.C.: DA, Comptroller of the Army), *USAF Cost and Planning Factors Guide* (AFP 173-13), et al.) present the standard unit, equipment, personnel and weapons systems cost factors used to calculate resource requirements of units and unit equivalents and the cost of operations. They are extremely useful when doing advanced research dealing with alternative force structures and cost variables.

2. The Annual Cycle: The defense and federal budgets are continually moving through stages of preparation or review, either in the executive branch or the Congress. There is, however, some similarity in the budget cycle every year regarding the stage of the budget and the documents available at a

TABLE 22
DEFENSE BUDGET CYCLE

Date	Action	Documents
Jan. (15 days after Congress convenes)	President submits budget, delivers State of the Union address	OMB Budget documents
End of Jan. (through April or May regularly)	Defense officials begin testimony before committees in support of budget	Posture statements, printed testimony, DOD Budget documents
Mar. 15	Committees submit preliminary report to Budget Committees	
Apr. 1	CBO submits reports	CBO Reports
Apr. 15	First concurrent resolution on the Budget	Budget Committee Reports
May 15	Concurrent resolution clears floor	
May-June	Committees report authorizations and appropriations approval	Committee Reports
May-Sept.	Mark-up and revision of Defense Budget, DOD input	
Sept. (7th day after Labor Day)	Congress completes action on new budget authority and spending authority	
Aug.-Sept.	Conference committee reports on Budget, laws authorizing spending are passed for Defense areas	Conference Reports, Public Laws
Sept. 15	Second concurrent resolution	
Sept.		Hearings become available
Oct. 1	Fiscal Year Begins	
Oct.-Jan.	Work continues on unfinished business, to include override or changes due to vetoes and completion of appropriations details	Public Laws

given time. Table 22 is a timetable of the budget process for any given fiscal year. It reflects the normal cycle of activity that is public and observable by a researcher. One must remember that at the same time the upcoming fiscal year's budget is being debated and worked on in Congress, the next budget in succession is being prepared in the Defense Department and the executive branch. Normally, at the beginning of the budget process—at the unveiling—there is a flurry of activity and writing in many magazines detailing the features of the Defense Budget and its strengths and weaknesses from many perspectives. These analyses should be consulted by the researcher interested in the general nature of spending. Those desiring to do in-depth research into the line item expenditures should consult the following referenced publications.

The Office of Management and Budget (OMB) coordinates the overall federal budget and publishes the budget material at the beginning of each year. The annual publications (all available from the GPO and at government depository libraries) are:

The Budget of the United States Government
The Budget of the United States Government—Appendix
Special Analyses—Budget of the United States Government
Mid-Session Review—Budget of the United States Government

These publications provide an overview of the federal budget, an explanation of spending programs, a description of the budget system, highlights of specified programs, economic and budget forecasts and histories and detailed agency documents. A researcher may want to get only the extracts of the budget dealing with the Defense Department, which is published as *The Budget of the United States Government and Budget Amendments: The Department of Defense Extract for Fiscal Year 19__*.

Supporting the OMB documents on the budget are the more specific and detailed DOD documents. An annual news release, "FY 19__ Department of Defense Budget" (Wash., D.C.: DOD, ASD(PA)), briefly describes the thrust of the Defense Budget and some of its provisions and programs. This general description is accompanied by the more detailed Posture Statements of the Secretary, Joint Chiefs of Staff and services mentioned elsewhere, and detailed accounting and statistical data on each one of the appropriations areas. Table 23 lists the standard appropriations titles in the Defense Budget. Each appropriations area has a set of generally well-known documents detailing spending in that area. These are discussed as follows:

TABLE 23
APPROPRIATION TITLES WITHIN THE DEFENSE BUDGET

Military Personnel:
Military Personnel, Air Force
Military Personnel, Army
Military Personnel, Marine Corps
Military Personnel, Navy
Reserve Personnel, Air Force
Reserve Personnel, Army
Reserve Personnel, Marine Corps
Reserve Personnel, Navy
National Guard Personnel, Air Force
National Guard Personnel, Army
Retired Pay, Defense

Research, Development, Test and Evaluation:
RDT & E, Defense Agencies
RDT & E, Air Force
RDT & E, Army
RDT & E, Navy
Director of Test and Evaluation, Defense

Operations and Maintenance:
O & M, Defense Agencies
O & M, Air Force
O & M, Army
O & M, Marine Corps
O & M, Navy
O & M, Air Force Reserve
O & M, Army Reserve
O & M, Marine Corps Reserve
O & M, Navy Reserve
O & M, Air National Guard
O & M, Army National Guard
Rifle Practice, Army
Claims, Defense
Contingencies, Defense
Court of Military Appeals, Defense

Procurement:
Aircraft Procurement, Air Force
Aircraft Procurement, Army
Aircraft Procurement, Navy
Missile Procurement, Air Force
Missile Procurement, Army
Other Procurement, Air Force
Other Procurement, Army
Other Procurement, Navy
Procurement, Defense Agencies
Procurement, Marine Corps
Procurement of Ammunition, Army
Procurement of Weapons and Tracked Combat Vehicles, Army
Shipbuilding and Conversion, Navy
Weapons Procurement, Navy

Military Construction:
Milcon, Defense Agencies
Milcon, Air Force
Milcon, Army
Milcon, Navy
Milcon, Air Force Reserve
Milcon, Army Reserve
Milcon, Naval Reserve
Milcon, Air National Guard
Milcon, Army National Guard

Family Housing:
Family Housing, Construction
Family Housing, Debt Payment
Family Housing, Operations
Homeowners Assistance Funds, Defense
Special Foreign Currency Program

Military Personnel (see Section IVE)
Research, Development, Test & Evaluation (see Section IVD5)
Procurement (see Section IVD3-4)
Military Construction (see Section IVF3)

Each of the appropriations areas also has a "Congressional Justification Book" prepared by the service appropriations area Comptroller which is furnished to the Congressional committees. These justification books are not normally printed in sufficient quantity to allow purchase by individual reseachers, but are unclassified and available for inspection at the Defense Department. The justification material normally includes line item accounting of funding requests and back-up historical and organizational or descriptive statistics and explanation to show the rationale behind requested funding. Portions of the justification material are often inserted into the hearings on the defense budget, but to a large extent most of this material goes unpublished.

A number of statistical reports compiled by the Defense Department provide a great deal of current and historical data on defense spending. *National Defense Budget Estimates for FY 19 __* (Wash., D.C.: DOD, ASD (C)) is an annual comprehensive presentation of historical, current and projected statistics and descriptive material on DOD spending, appropriations, programs and pay. *Financial Summary Tables: Department of Defense Budget for Fiscal Year 19__* (Wash., D.C.: DOD, ASD(C)) is another annual publication which presents statistical tables on the defense budget, including obligations (total obligational authority), appropriations, budget authority and tradeoffs. Two Defense Department statistical publications account for money spent and actually appropriated, as opposed to supporting the budget submission. The quarterly report *Military Functions and Military Assistance, Status of Funds* (Wash., D.C.: DOD, ASD (C)) presents cumulative data (by month) on DOD outlays and obligations with the 30 September issue containing the fiscal year cumulative and comparative with preceding fiscal years. (This publication was formerly known as *Status of Funds by Functional Title.*) *Congressional Action on DOD Authorization and Appropriation Requests by Appropriation Account and Item, FY__* (Wash., D.C.: DOD, OASD (C)) is an annually prepared report which traces Congressional action on Defense requests through the authorization, appropriation, conference committee, floor action, DOD appeal and final appropriation stages, showing Congressional increases and decreases after the complete budget cycle. This is a particularly useful document for analyzing both Congressional support and opposition for certain programs.

The publications mentioned so far are the OMB and DOD documents which support the defense budget. Once the budget is presented and goes into the hearings stage, a number of legislative branch publications of importance to the researcher become available. The Congressional Budget Office submits two annual reports (and a number of ad hoc special analytical studies on specific features of the budget) of interest: *Five-Year Budget Projections and Alternative Budgetary Strategies for Fiscal Years 19__-19__: A Report to the Senate and House Committee on the Budget* (CBO, Wash., DC: GPO) (previously known as *Five-Year Budget Projections*) and *An Analysis of the President's Budgetary Proposals for Fiscal Year 19__* (CBO, Wash., D.C.: GPO) (previously known as *Overview of the 19__ Budget*). The Congressional Research Service also compiles an Issue Brief, *Defense Budget, FY__* annually.

Finally, the printed Hearings (see Table 5), Reports, and the actual Public Laws become available to the researcher at the end of the annual cycle, and are essential resources. Legislative histories compiled for the various military-related acts (Defense Appropriations, Military Construction, Foreign Assistance, et al.) are normally compiled by large law libraries and include all the documents relevant to the history of a public law. *Public Laws Legislative Histories* (Chicago, IL: CCH, Inc., 1979-) is a new microfiche service of Commerce Clearing House which includes up-to-date legislative histories of public laws as they are passed, including the House or Senate bill as introduced, the reported House or Senate bills, Committee Reports, Conference Committee Reports, and relevant legislative debate in the *Congressional Record*. This is a very useful service for current research, with the printings lagging only a few months behind the passage of laws. *Public Laws Legislative Histories* does not include reproduction of the hearings, but these can be traced using *CIS Abstracts/Index*. The General Accounting Office, which compiles its own internally used legislative histories, is presently preparing to make these available through the depository library system, which will provide great accessibility to researchers. Another common and useful source for tracking public laws and their legislative histories is *U.S. Code Congressional and Administrative News* (St. Paul, MN: West Publishing Co.), a monthly service reporting new laws (including their text) with summaries and legislative histories (sources listed without texts). This service is normally a few months ahead of *Public Laws Legislative Histories*.

D. THE MILITARY-INDUSTRIAL COMPLEX

In the first three sections of this part, sources dealing with the internal institutions of the U.S. military have been discussed. The many companies and institutions that support the U.S. military through research and development, production of materials, or contracting for services—the "Military-Industrial Complex"—are also the subject of a great deal of research interest, and are discussed in this section.

First, introductory sources for doing business research on the nature of the corporate institutions and their activities are discussed. Then descriptive material detailing the actual participants are presented. Finally, sources on DOD procurement and contracting policy are presented, together with materials on the actual current research and development and procurement programs of the military.

A great deal has been written on the military-industrial complex, some of the works the product of excellent primary research. This section attempts to represent the range of materials which enable the researcher to understand the dynamics of the complex itself, its role and position in the U.S. economy and society, and its current activities. A general and selective unannotated bibliography of some works on the military-industrial complex is *The Military Industrial Complex: A Source Guide to the Issues of Defense Spending and Policy Control* (Thomas A. Meeker, Los Angeles, CA: Center for the Study of Armament and Disarmament, CSU, 1973).

1. Researching the Complex and Companies: Business research is a specialized field with a number of research guides. *Where to Find Business Information* (David M. Brownstone and Gorton Carruth, New York: John Wiley, 1979) is the most up-to-date and useful guide for quick reference on worldwide references and information services. Its subject and publisher index is comprehensive and relevant to military-related businesses. *Business References Sources* (Lorna M. Daniels (Comp.), Cambridge, MA: Baker Library, Graduate School of Business Administration, Harvard University, 1971) and *Sources of Business Information* (Edwin T. Coman, Jr., Berkeley, CA: Univ. of California Press, 1970) are two general sources, which describe the sources for finding facts and statistical data about businesses and industries. *How to Find Information about Companies* (Wash., D.C.: Washington Researchers, 1979) is not as comprehensive as the sources mentioned above but contains extensive information on using the resources of the federal government agencies and the Freedom of Information Act in getting information.

The corporate world also has its own periodicals and trade journals which are great resources for information on corporate programs and activities. Three indexes serve as research tools for information about companies, industry groups, activities, contracts and developments. *Business Periodicals Index* (New York: H.W. Wilson Co., 1913-1958-) is a widely available monthly index to approximately 200 periodicals in finance, business and related fields, including listings by company and corporation. *F&S Index of Corporations and Industries* and *F&S International Index* (Cleveland, OH: Predicasts, Inc.) are indexes of articles from financial, trade and business publications about companies and industry groups. These indexes are invaluable for following the activities and contracts of a company. *The Wall Street Journal Index* is also valuable for business research. Finally, there is an index of magazines which are produced by corporations for public relations or for internal information. Many of the aerospace and large defense industries produce these magazines, some of which are both informative and revealing. *Gebbie's House Magazine Directory* (Burlington, IA: National Research Bureau, 1974) is a directory of various house organs.

The business research guides and the magazine indexes are the primary tools for finding information about companies and activities and developments in industry. Information on companies active in the military-industrial complex (including their work and contracts) can be obtained from organizations which study the complex, such as DMS, Inc., Frost & Sullivan, and the Council for Economic Priorities. Other basic sources of descriptive data on companies and corporations are some reference books that are widely available in most libraries. The research guides described above discuss a number of other sources, but these books are the basic and most authoritative sources:

 American Register of Exporters and Importers (annual)
 Directory of Firms Operating in Foreign Countries
 Directory of Corporate Affiliations (annual, with five supplements)
 Dun & Bradstreet Middle Market Directory (annual)
 Dun & Bradstreet Million Dollar Directory (annual)
 Dun & Bradstreet Reference Book (revised bimonthly)
 Moody's Industrial Manual (annual)
 Standard and Poors Corporate Descriptions (biweekly service)
 Thomas Register of American Manufacturers (annual)

Who Owns Whom (annual)

Other descriptive sources are also worth noting. Information on the 11,000 or so public companies (those selling stock to the public) in the United States (most defense prime and sub-contractors are in this category) can be obtained from the Securities and Exchange Commission (SEC). An SEC publication *The Directory of Companies Filing Annual Reports with the Securities and Exchange Commission* (Wash., D.C.: GPO) is an annual listing of these companies. The information available from the SEC, primarily organizational and financial papers, can be written for or inspected at SEC public reading rooms in Chicago, Los Angeles, New York and Washington, D.C. Copies of the different forms required to be filed by these companies with the SEC can be purchased by writing:

Securities and Exchange Commission
Public Reference Section
Washington, D.C. 20549
(202) 528-5360

The public companies, which also issue glossy and informative annual reports, can also provide much information.

Information about nonprofit organizations (which includes many think tanks and associations) in the form of U.S. income tax returns (Form 990) is available from the Internal Revenue Service. A copy of these forms is available either in person or by writing:

Internal Revenue Service
FOI Reading Room
1111 Constitution Ave., N.W.
Washington, D.C. 20224
(202) 566-3700

The *Cumulative List of Organizations, Publication No. 78, Described in Section 170(C) of the Internal Revenue Code of 1954, Revised to October 31, 1978* (IRS, Wash., D.C.: GPO) lists these organizations and is available from the GPO by subscription (annual with three quarterly cumulative supplements).

2. Defense Industries and Contractors: Almost every industry in the United States is a prime or sub-contractor with the Defense Department, or has some continuing or contingency relationship with the military. There are a number of reference works which list these industries and include the amount of their contracts. The major source of information on organizations which do work directly for the military is the DOD-produced Prime Contractor series, which breaks down contractors by amount, category and location. There are a number of different titles in the series, produced annually and semiannually by the Washington Headquarters Service of the Office of the Secretary of Defense. Most are listed in *Catalog of DIOR Reports* (Wash., D.C.: DOD, WHS(DIOR), 1980):

Educational and Nonprofit Institutions Receiving Military Prime Contract Awards for Research, Development, Test & Evaluation (annual)

Prime Contract Awards in Labor Surplus Areas (annual)

Military Prime Contract Awards, Size Distribution (annual)

Military Prime Contract Awards by State for FY 1951 to Date (annual)

Prime Contract Awards by State, ___ half Fiscal Year 19__ (semiannual)

Military Prime Contract Awards, Fiscal Year __ (annual)

Military Prime Contract Awards, Oct 19 __ - Mar 19 __ (semiannual)

Military Prime Contract Awards by Region and State, Fiscal Years 19__, 19__, 19__ (annual)

Military Prime Contract Awards, by State, Fiscal Year 19__ (annual)

Department of Defense, State Distribution of Military Prime Contract Awards Including Five Leading Contractors and their Major Work (semiannual)

DOD Prime Contractors which Received Awards of $10,000 or More (annual)

100 Companies, Companies Receiving the Largest Dollar Volume of Military Prime Contract Awards, FY__ (annual)

500 Companies, Contractors Receiving the Largest Dollar Volume of Military Prime Contract Awards for RDT&E, FY__ (annual)

Procurement for Small or Other Business Firms (monthly)

Military Prime Contract Awards by Service Category and Federal Supply Classification: Fiscal Years 19__, 19__, 19__, 19__ (annual)

Complete data for contract awards, by location of contract, contractor, and contractee, is compiled by the DOD in two annual publications of over 1,000 pages each:

Military Prime Contract Awards by State, City, and Contrac-

tor (Wash., D.C.: DOD, WHS(DIOR), dollar amount of awards by contract, city, and state, including foreign contracts.

Military Prime Contract Awards over $10,000 by State, County, Contractor, and Place (Wash., D.C.: DOD, WHS (DIOR)), state, city, and county contract awards listing by contractor and awarding department.

The combined information in both of the above publications is also available separately for each state. Two subcontracting publications are also compiled by DOD:

Companies Participating in the DOD Subcontracting Programs (Wash., D.C.: DOD, WHS(DIOR)), a quarterly report on approximately 650 companies which are military subcontractors.

Geographic Distribution of Subcontracting Awards (Wash., D.C.: DOD, WHS(DIOR)), compilation by state and by prime contractor.

Besides the Prime Contractors series, there are a number of privately produced listings of Defense contractors and industry which identify and quantify their participation. The best of these is the *Defense Industry Organization Service* (Wash., D.C.: Carroll Publishing Co.), a semiannually updated directory of top companies including organization charts and listings of key personnel. *Aerospace Facts and Figures 19__/__* (Aerospace Industries Assn. of America, New York: Aviation Week and Space Technology) is another good source, presenting a statistic and thematic survey of the aerospace industry with information on production, missile and space programs, R&D, foreign trade and employment.

Three other works which list federal contractors and present supplemental information are valuable. *Directory of Federal Contractors* (Wash., D.C.: Washington Research Service, 1979) is a listing and directory; *Government Production Prime Contractors Directory* (Wash., D.C.: Government Data Publications, 1978) is an alphabetic and zip code listing with full addresses; and *U.S. Contract Awards* (Wash., D.C.: Washington Representative Services, 1979) is a comprehensive listing of prime contractors in five volumes by agency, company, public sector and procurement area.

The *Register of Planned Emergency Producers, Vol. I: Alphabetical Listing, Vol. II: Geographical Listing* (DOD 4005.3H) (DOD, OASD (MRAL), Wash., D.C.: GPO, 1977) lists geographically and alphabetically industrial firms participating in the DOD Industrial Preparedness Program of planned producers of materiel in time of war. It also lists DOD owned or operated plants which assemble or produce military materiel. This listing is a good indication of the number of companies which would support the military in a war.

Finally, defense industries and contractors can be identified by the numerous associations which promote defense spending and act as fraternities of like industries for lobbying and information services. Table 24 lists some of the largest defense-oriented associations and their publications. These organizations also are fraternities for specialists in the military and industry in these fields. Usually a list of corporate sponsors or members appears in the magazines. In addition, many of the associations publish listings of corporate members.

3. *Contracting and Procurement Policy:* Selling to the Department of Defense and contracting with the military is controlled by a series of regulations which are available for any researcher to examine. These various regulations are revealing as to the method by which contractors are selected, and what standards they are assumed to follow. There is no good reference work which explains how the contracting system works. *Selling to the Military* (DOD, Wash., D.C.: GPO, 1979) is a simple introduction to government buying, research & development, procure-

---

TABLE 24
DEFENSE ASSOCIATIONS AND PUBLICATIONS

Air Force Association
  Air Force Magazine
American Defense Preparedness
  Association
  National Defense
Armed Forces Communications
  and Electronics Association
  Signal Magazine
Army Aviation Association
  of America, Inc.
  Army Aviation Magazine
Association of the U.S. Army
  Army Magazine
Marine Corps Association
  Leatherneck
  Marine Corps Gazette

National Defense Transportation
  Association
  Defense Transportation Journal
National Guard Association of
  the U.S.
  National Guard
Navy League of the U.S.
  Seapower
Reserve Officer's Association
  of the U.S.
  The Officer
Retired Officer's Association
  The Retired Officer
United States Naval Institute
  USNI Proceedings

ment of commodities and supplies and providing services to the military.

The basic regulation on military procurement policy is the *Defense Acquisitions Regulations* (formerly the *Armed Services Procurement Regulations*) (DOD, Wash., D.C.: GPO). The DAR establishes DOD policies relating to procurement, and purchases and contracts for supplies and services. The basic manual is updated by *Defense Procurement Circulars* and additional supplements and clarification. Each one of the services has implementing regulations of the DARs. *Army Procurement Procedures* (DA, Wash., D.C.: GPO, 1976-), *Air Force DAR Supplement* (DAF, Wash., D.C.: GPO, 1974-), and *Navy Contracting Directives* (DN, Wash., D.C.: GPO, 1978-) (formerly *Navy Procurement Directives*) outline uniform policy and are available on a subscription basis from the GPO. The *Defense Supply Procurement Regulations* (DLA, Wash., D.C.: GPO, 1976-) provides implementation for the Defense Logistics Agency.

The Commerce Clearing House has three information services which deal with federal and DOD contracting and provide much background information on policies and programs. *Cost Accounting Standards Guide* (Chicago, IL: CCH, Inc.) is a loose-leaf service with monthly updates of cost accounting standards required of defense contractors. The *Board of Contract Appeals Decisions* (Chicago, IL: CCH, Inc.) is a two-volume updated directory of decisions of the Armed Services Board of Contract Appeals and other Appeals boards of the U.S. government. *Government Contracts Reports* (Chicago, IL: CCH, Inc.) consists of eight base volumes with weekly updates of laws and regulations dealing with government contract work, including defense production, procurement, standards and costs.

The *Defense Contract Audit Manual* (DCAA, Wash., D.C.: GPO) is another regulatory document covering Defense auditing procedures of contract awards.

4. Current Procurement: The current performance of the defense industries is detailed in a number of publications. The prime contracts awards series mentioned above give much information on the largest industries in the military-industrial complex, and there are numerous other sources to track and keep abreast of current procurement and the status of defense production. *Commerce Business Daily (A Daily List of U.S. Government Procurement Invitations, Contract Awards, Subcontracting Leads, Sales of Surplus Property and Foreign Business Opportunities)* (Dept. of Commerce, Wash., D.C.: GPO) is the major source of current government needs in the way of services, supplies, equipment and material. Together with the daily Defense Department News Release, "Contract Awards," this covers most of the current prime contract information on the status of procurement and contracting.

The current procurement programs of the Defense Department are examined annually in the defense budget hearings. The major weapons procurement programs are normally analyzed in great detail. A number of annual Defense Department reports are produced dealing with the procurement portions of the budget. *Program Acquisition Costs by Weapon System: DOD Budget, FY__* (Wash., D.C.: DOD) gives the fiscal year acquisition and shows the requested funding for procurement, R&D and military construction associated with the major systems. The total procurement portions of the budget are outlined in two publications: *Procurement Program: DOD Budget for Fiscal Year 19__* (Wash., D.C.: DOD) and *Procurement Programs (P-1), Department of Defense Budget for Fiscal Year 19__* (Wash., D.C.: DOD). *Procurement Program* is a pocket book listing the line items in the fiscal year budget and the amount and quantity requested. *Procurement Programs (P-1)* is a four-year trace of DOD-wide procurement also by line item but showing the last two fiscal years and the next planned fiscal year costs, quantities, and unit costs for the current submission. More detailed information on each one of the line items in each procurement title of the appropriations bill (see Table 23) is contained in the justification material submitted to Congress along with the budget. This material is compiled by the appropriations comptroller. The Hearings also reproduce much of this justification material. The CRS has produced a number of Issue Briefs on current DOD procurement programs including analysis of the Trident, Cruise Missile, MX, V/STOL, and various Fighter Aircraft programs. The GAO also issues reports on current procurement programs of the DOD.

Detailed data on the financial status of the major procurement programs is contained in three publications. *Funding Status of Major Weapon Systems* (Wash., D.C.: DOD, WHS (DIO&R)) presents semiannual statistics. *SAR Program Acquisition Cost Summary* (Wash., D.C.: DOD) is a quarterly summary of current program information and cost estimates from *Selected Acquisition Reports* on approximately 40 major weapon systems, and *Financial Status of Selected Major Weapons Systems* (Wash., D.C.: GAO) is a semiannual accounting and acquisition audit.

5. Research & Development: Military research and development (R&D) is undertaken mostly by defense industry, although some is done internally by the Defense Department. Contract awards for research, development, test and evaluation (RDT&E) are reported with other contract award sources already cited. One can follow military R&D by following these sources, reading the military magazines or the military newsletters.

Background information on military R&D is contained in two publications: *RDT&E Management Guide* (NAVSO P-2457) (Navy, Wash., D.C.: GPO, Dec. 1979) and *An Inventory of Congressional Concern with Research and Development* (Report of the Subcommittee on Government Research, Committee on Government Affairs, U.S. Senate, Wash., D.C.: GPO, 1966-). The *RDT&E Management Guide* is an excellent reference on DOD and Navy R&D management, organization and policy. *An Inventory of Congressional Concern* is a continuing bibliography of congressional publications dealing with science and technology and research and development.

The research and development community is a group of scientific laboratories conducting basic scientific research, and other organizations working on research projects to develop weapons systems. A number of sources list various research organizations and describe their work. Three directories list numerous types of research organizations (laboratories, industry, federal in-house, universities) and briefly describe their work:

Directory of Federal R&D Installations (NSF, Wash., D.C.: GPO, 1970?-)

Federal Laboratory Consortium Resource Directory (NSF, Wash., D.C.: GPO, September 1978)

Research Centers Directory (Detroit, MI: Gale Research Co., 1963?-)

Another directory is *The University-Military-Police Complex: A Directory and Related Documents* (Michael T. Klare, comp., New York: NACLA, 1970), an out-of-date description of the activities of university and non-profit research organizations contracting with the Defense Department, but still a valuable listing since most are still large defense contractors.

Current research and development programs are outlined in detail every year as part of the defense budget hearings. A number of reports are issued by the Defense Department annually and much information on research and development programs is contained in the research and development portions of the printed hearings. The overall DOD program of research and development is presented in *The FY __ Department of*

Defense Program for Research, Development and Acquisition (Wash., D.C.: DOD, USDRE, 1978-), a bound publication of the testimony for the record of the Chief of DOD R&D programs. This report is supplemented by the annual report of the Defense Advanced Research Projects Agency (DARPA)—*Defense Advanced Research Projects Agency, Fiscal Year 19__ Program for Research and Development* (Wash., D.C.: DARPA, DOD)—a description of advanced and high risk programs, accomplishments and applications. DARPA programs are often the high technology programs which result in important breakthroughs in military technology. These two reports describe the various R&D programs of the Department of Defense overall. Each of the services also has an R&D report, and printed and bound versions of congressional testimony, which discuss in more detail the rationale behind military programs. The Army report, *Equipping the Army of the Eighties: A Statement to Congress on the FY__ Army RDT&E and Procurement Operations* (Wash., D,C.: DA, 1978-), is the most readable of the reports and provides much information on the U.S. and Soviet ground warfare programs. Finally, the Defense Department activities in space are outlined in *Department of Defense Activities in Space and Aeronautics, FY__* (Wash., D.C.: USDRE, DOD), a bound copy of the annual testimony before the Subcommittee on Science, Technology and Space of the Senate Committee on Commerce. This report reviews DOD's participation in the Space Shuttle and other space programs.

The actual line items in the Research and Development portions of the budget are listed in the pocket book *RDT&E Program: DOD Budget for Fiscal Year 19__*(Wash., D.C.: DOD) which lists program element titles and the requested current funding. *RDT&E Program (R-1): Department of Defense Budget for Fiscal Year 19__* (Wash., D.C.: DOD) breaks down these categories into more detail and shows a four-year trace of expenditures including the projected fiscal year. The justification material submitted to Congress describes each program element in detail. The *Justification of Estimates for Fiscal Year 19__ Submitted to Congress: RDT&E, Defense Agencies, Director of T&E, Defense* (Wash., D.C.: DOD, OSD) covers the OSD/JCS and the Defense agencies including budget, personnel and programs. This book is available for sale from the DOD Freedom of Information office. The justification material for the other appropriations titles dealing with research and development (see Table 23) are less accessible but important for serious and advanced research.

E. MILITARY PERSONNEL

Personnel issues should not be overlooked when considering military and strategic affairs. Indeed, today the greatest internal problems facing the U.S. military are personnel problems. Military manpower policy and programs are partial causes and solutions to manpower problems but overall national policy vis-a-vis an all-volunteer versus a drafted force have great impacts on internal personnel programs and cohesion. Therefore, when researching personnel issues one must consider national policy as reflected in laws and executive prerogative, internal personnel policy, the interrelationship between forces and weapons and personnel and demographic indicators impacting on the make-up of the military force. Each of these areas are well-documented and thus this area is ripe for research.

1. Policy, Costs and Compensation: Manpower strength levels and personnel programs are covered in detail in the annual hearings on the defense budget. At least one volume of each of the four sets of hearings is devoted to the strength and status of military personnel and civilian employees of the Department of Defense. These hearings give the state of the personnel of the military and discuss programs, policy and strengths. Personnel policy is outlined in internal regulations of the Defense Department and are traceable through the indexes discussed under Section IIIC4. A reliable compilation of pay and benefits information and other information dealing with personnel policy is the *Uniformed Services Almanac, 19__* (Wash., D.C.: Uniformed Services Almanac), an annual handbook directed at military personnel. A number of commercial books directed at military personnel (e.g., *The Army Officers Guide*) also provide a tremendous amount of information on personnel policy. These publications are discussed in Section IVA. One doing advanced research on internal personnel policy should also be familiar with the many personnel newsletters produced by the various personnel offices:

> *Air Force Civilian Personnel Newsletter*
> *Air Force Officer's Career Newsletter*
> *Army Personnel Letter*
> *Career Advisory News* (USAF)
> *Careergram* (USN)
> *Officers' Personnel Newsletter*
> *Manpower and Organization Newsletter*
> *Persfacts* (USAF)

A number of military periodicals also follow personnel policy, benefits, management and issues. *Air Force Times, Army Times* and *Navy Times* most regularly report personnel issues. The largest military magazines—*Air Force Magazine, Army, Leatherneck, Marine Corps Gazette, USNI Proceedings*—also cover personnel issues. Two other journals—*Armed Forces and Society* and *Journal of Political and Military Sociology*—cover military personnel issues and policy, military sociological and demographic issues and civil-military relations.

Studies of military manpower policy and costs are divided into three categories: official commissions and study groups, annual military manpower reports, and studies of the all-volunteer force and the draft. Annual military manpower reports and studies of the All Volunteer Force are covered in the next sections. Official manpower studies have been conducted for a number of issues including compensation, recruiting, training, mobilization and utilization. The grandfather of all recent manpower studies is the Defense Manpower Commission examination of military manpower in the mid-seventies. *Defense Manpower: The Keystone of National Security* (Defense Manpower Commission Report to the President and the Congress, Wash., D.C.: GPO, Apr. 1976) is the final report of the Commission and together with the staff studies and supporting papers (printed separately), this commission is the most useful recent study of overall manpower policy and compensation available to the researcher. Other recent studies of manpower policy and costs are available as Congressional hearings and publications. The Congressional Research Service has done a number of issue briefs and reports on manpower policy, mobilization manpower, and manpower costs. The Congressional Budget Office also conducted two studies for the FY77 Defense Budget: *The Costs of Defense Manpower: Issues for 1977* (CBO, Wash., D.C.: GPO, Jan. 1977) and *Defense Manpower: Compensation Issues for Fiscal Year 1977* (CBO, Wash., D.C.: GPO, Apr. 1976). Hearings have also been held relating specifically to manpower: hearings on unionization of the military services held by the Armed Services Committees in 1977 and hearings on military recruiting practices held by the Armed Services Committees in 1977 and 1978. More recently hearings have been held on the All-Volunteer Force.

Finally, there are two DOD annual reports on manpower policy and programs which are essential reference aids. These reports support DOD requests for specific manpower levels as part of the annual authorization hearings. *Manpower Requirements Report for FY__* (Wash., D.C.: OASD (MRAL), 1973-)

explains and justifies strengths requests for the coming fiscal year and presents a breakdown of assignment of military manpower by programs and combat or support areas. Special annexes to the report talk about issues such as the cost of manpower, manpower in the Defense agencies and security assistance functions and women in the military. An annex to the *Manpower Requirements Report* is *Military Manpower Training Report* (Wash., D.C.: OASD (MRAL), 1973-), which justifies requests for funding of training program loads, student and full-time faculty and staff levels at training bases and military educational institutions.

2. Manpower Strength Figures: Military manpower and civilian strengths is of great interest and a number of specific sources of data present both historical and contemporary information. An annual publication of the Defense Department, *Selected Manpower Statistics* (Wash., D.C.: DOD, WHS (DIO&R)) presents statistical data on active military, civilian, reserve, and retired personnel incuding historical information. Each chart is updated quarterly, some monthly, and most are available on request from the DOD Public Affairs Office. *Military Manpower Statistics as of__* (Wash., D.C.: DOD, WHS (DIO&R)) and *Civilian Manpower Statistics* (Wash., D.C.: DOD, WHS (DIOR) are monthly statistical breakdowns of similar statistics.

Current statistics are available in the form of updated charts from *Selected Manpower Statistics* or from other recurring charts made available to the public. "Military Strength Figures Summarized by DOD" (OASD (PA) News Release) is a monthly report on actual total military strength and authorized planned strength. "Military Manpower Strength Assessment and Active Force Recruiting Results" (OASD(PA) News Release) is a quarterly report on the strength, accessions, quality and characteristics of recruits compared with objectives and authorizations. "Department of Defense Active Duty Military Personnel Strengths by Regional Area and By Country" (OASD (PA), News Release) is another source issued quarterly and breaking down assignment of American military personnel stationed overseas by service, country and area. *Worldwide Manpower Distribution by Geographic Area* (Wash., D.C.: DOD, WHS (DIOR)) is a similar quarterly report including information on military and civilian personnel and dependents.

A number of other current sources of manpower information break down manpower by service, functions, assignments, etc.

Department of Defense Civilian and Military Personnel: OSD/ JCS and Other Defense Agencies (Wash., D.C.: DOD, WHS (DIO&R)) is a monthly accounting of personnel assigned to headquarters agencies. Each of the services has a much more specific publication which breaks down military and civilian strength and distribution. For instance, the Navy publishes *Navy Military Personnel Statistics* (NAVPERS 15658), a quarterly breakdown of personnel strength, accessions, and attrition, with the September issue containing data for the entire fiscal year.

Reserve component personnel statistics are presented in two publications: *Official Guard and Reserve Manpower: Strengths and Statistics* (Wash., D.C.: OASD(MRAL)) and *Reserve Forces Manpower Charts* (Wash., D.C.: OASD(MRAL)). *Official Guard and Reserve Manpower* is a monthly comprehensive presentation of statistics detailing reserve component personnel and force characteristics including state breakdowns of strength, personnel demographic profiles and gains and losses. *Reserve Forces Manpower Charts* is a quarterly chartbook of statistical data on the size and composition of the reserve components. Breakdowns of personnel by base and location are discussed in Section IVF3.

3. The All-Volunteer Armed Forces and the Draft: The subject of the All-Volunteer Force (AVF) overlaps all of the areas dealing with personnel policy and costs. It has been the subject of numerous hearings and congressional reports, private studies and Department of Defense studies and reports. Many studies on the draft, the issue of conscription and the feasibility of an AVF were written in the mid-1970s when the draft was abolished and the AVF was instituted. *Conscription: A Select and Annotated Bibliography* (Martin Anderson, ed., Stanford, CA: Hoover Institution Press, 1976) contains references to many of these works as well as sources on military manpower procurement and policy worldwide.

More recently, numerous hearings have been held on the status and cost of the present AVF. The House Armed Services Committee held hearings in 1978 and the Senate Armed Services Committee held hearings in March, 1977, and February and June, 1978. The Congressional Budget Office conducted two studies in 1978 at the height of the "success-of-the-AVF" debate—*The Selective Service System: Mobilization Capabilities and Options for Improvement* (CBO, Wash., D.C.: GPO, Nov. 1978) and *National Service Programs and their Effects on Military Manpower and Civilian Youth Programs* (CBO, Wash.,

D.C.: GPO, Jan. 1978)—both of which contain valuable background information and policy alternatives. The Congressional Research Service has also done a number of Issue Briefs and reports on the subject of AVF and manpower mobilization.

The primary DOD study on the AVF is *America's Volunteers: A Report on the All-Volunteer Armed Forces* (Wash., D.C.: DOD, OASD(MRAL), Dec. 31, 1978). This is a very optimistic appraisal of the adequacy of the AVF and a useful resource. An earlier report of the same office, *The All-Volunteer Force: Current Status and Prospects* (Wash., D.C.: DOD, OASD(MRA), 1976) is also useful. *Achieving America's Goals: National Service or the All-Volunteer Armed Force?* (Study Prepared by the Subcommittee on Manpower and Personnel of the Senate Armed Services Committee, 95-1, Wash., D.C.: GPO, 1977), another appraisal of the AVF at the end of 1976, considers policy alternatives and problem solving. A final source of information is a number of Rand Corporation studies and reports on military manpower and the All-Volunteer Force conducted for the Defense Department. These are traceable through *Selected Rand Abstracts.*

The more recent debate on the draft and registration has not produced much new research or reference material. Congress members on different sides of the issue have released many private studies backing their views, and organizations have sprung up or taken positions on the issues of registration, women and the military, the draft and the success or failure of the All-Volunteer Force.

4. Women in the Military: The issue of women in the military—their role and contribution—is of continuing interest and importance. Much has been written in the general and military media, ranging from unsubstantiated opinion to positive and negative testimonials, to uninformed scholarly works that are more review than policy or feminist-oriented. Two bibliographies dealing with women in the military are *Women in the Military* (USAF Academy, Colorado Springs, CO: 1975), a thorough retrospective bibliography and *Women: A Selective Bibliography* (Army Library, Wash., D.C.: 1975 (with supplements)), a broader bibliography including references to women in the military with annual supplements. A number of feminist organizations have also done research and work towards equity for women in the military. The Women's Equity Action League Education and Legal Defense Fund (805 15th St., N.W., Suite 822 Wash., D.C. 20005), for instance, has compiled an information kit, "Women and the Military," which includes issue papers and bibliographies.

The issue of women in the military has been dealt with annually as part of the personnel portions of the military budget hearings since 1973. The major Defense Department report on women in the military is *Use of Women in the Military* (Wash., D.C.: DOD, OASD(MRAL), May 1977 (Revised, Sept. 1978)), an examination of utilization and policy with statistical analysis. A section of the *Manpower Requirements Report* outlines DOD policy annually and should be consulted. The Congressional Research Service has also compiled an Issue Brief on Women in the Military. Another study that is often referenced is *Women in the Military* (Martin Binkin and Shirley Bach, Wash., D.C.: Brookings Institution, 1977), another statistical and policy-related analysis. Finally, "Women as New Manpower," (Anne Hoiberg, ed., in *Armed Forces and Society,* Summer 1978, Vol. IV, No. 4) is a more critical and feminist view of the issue of women's role in the military and contains ten essays and an annotated bibliography.

F. THE LOCAL IMPACT OF MILITARY PRESENCE AND SPENDING

Elements of the military or producers in the military-industrial complex are present and influential in every community of the UnitedStates. Military bases, reserve centers, armories, industries, university research centers, and contractors contribute a great deal of money to the local economy and are often the most important employers. Changes in defense posture or spending can affect a community in many serious ways by creating an infusion of money and military prosperity or by stopping a community's livelihood in the form of base closings or loss of contracts or business. In either case, the local impact of military presence is the tendency to form an opinion of the military on the basis of the payoff to the community.

Researchers, legislators and activists may therefore want to quantify and describe the nature of the military presence in the community and develop a data base to judge its influence. Most of the information needed is available from a number of government sources described in this section and in the previous section, **The Military-Industrial Complex.** The Council on Economic Priorities, a non-profit research organization in New York, specializes in the issue of the local impact of military spending. Besides doing research in the issues of defense dependency, conversion and the effects of defense spending, the Council provides a free reference service. The Council's publications—*CEP Studies, Reports, Newsletter* and *CIC Update*—detail various aspects of military production, employment, influence, dependency, basing and conversion.

1. Factors Affecting the Community: The best description of parameters for a methodology for quantifying the extent of military and industrial influence and involvement in local areas is "The Military Industrial Complex in Iowa," (Alvin R. Sunseri, in *War, Business and American Society,* Benjamin Franklin Cooling, ed., Port Washington, NY: Kennikat Press, 1977), which examines all of the different aspects of the military presence in a state with a seemingly nonexistent military presence. A number of studies have examined the extent of military involvement in specific states, including *The Dark Side of Paradise: Hawaii in a Nuclear World* (Honolulu, HI: Catholic Action of Hawaii/Peace Education Project, Sept. 1980), *New Mexico, the Military, and the Bomb* (Albuquerque, NM: New Mexico People & Energy, 1980), and *The Military Presence on the Gulf Coast* (William S. Coker, ed., Pensacola, FL: Gulf Coast History and Humanities Conference, 1977). Another introductory source is *The Impact of Defense Establishments on Communities: A Bibliography* (William J. Low, Vance Bibliographies, P.O. Box 229, Monticello, IL 61856, Dec. 1978).

Three studies also give information on the local impact of military spending. *The Pentagon Tax: The Impact of the Military Budget on Major American Cities* (Dr. James R. Anderson, Lansing, MI: Employment Research Associates, March 1979), is the latest of an annual series of reports which uses data on taxes and expenditures. *The Empty Pork Barrel: Unemployment and the Pentagon Budget* (1978 ed.) (Marion Anderson, 590 Hollister Bldg., Lansing, MI 48933 (Processed)) is a study of a similar type using similar data. *Fiscal Year 19___ Geographic Dissemination of Federal Funds* (Wash., D.C.: Community Services Administration) is the granddaddy of reports containing statistics in many areas breaking down federal spending by state. It is the source of much raw data.

Bases and military commands are also sources of information on military spending and presence. Military installations can provide information on manpower and employment, pay, local contracting for services, procurement of commodities and the mission and responsibilities of units and activities at military installations. Many bases produce "stockholder reports" which provide much of this information and a description of the base. Local information sources also provide data on employment, defense-related industry, procurement and military units. Besides the public information offices of the military installations and the defense industries, one should go to the local Chamber of Commerce, libraries, and newspapers.

2. Sources of Comparative Data: Besides *Geographic Distribution of Federal Funds* already mentioned and the previous section on the military-industrial complex, there are a number of sources a researcher could pursue for comparative data on the extent of military involvement in the states.

A number of state publications provide helpful information. Every state has a "Blue Book" or "Legislative Manual" which often provides data relating to the active military and reserve components in the state, specifically the state National Guard. Two indexes of state publications which can also lead to material describing employment, pay and other general statistics on the state are *Monthly Checklist of State Publications* (Wash., D.C.: Library of Congress, 1910-), a state-by-state monthly listing with annual index, and *State Manuals, Blue Books and Election Results* (Charles Press and Oliver Williams, Berkeley, CA: Univ. of California, 1962). The *Encyclopedia of Geographic Information Sources* (see Section IVD1) is also an excellent source of references and statistical works broken down by state including yearbooks, directories, periodicals, indexes and statistics.

Information on Congressional Districts is also of interest because of the political connection between government spending and localities. Maps of Congressional Districts appear in the *Congressional Directory* and *The Almanac of American Politics* (Barone, Ujifusa and Matthews, Boston, MA: Gambit, Inc., 1972-) which give some insights into the traits of each district. The basic source of information by Congressional District is *Congressional Districts in the 1970s* (2d Ed.) (Wash., D.C.: Congressional Quarterly, 1974), which contains demographic and statistical information broken down by district and information on military, education, industrial and communications facilities in each district.

Other sources of comparative information are specific sources on military bases (discussed in the next section), military procurement and prime contracts (discussed in Section IVD3-4) and military and civilian personnel (see Section IVE2).

3. Military Bases: Military Bases and installations in the

United States (overseas bases are covered in Section IVG) are the most visible indications of the extent of military presence and influence. Excellent government records are kept on its property holdings and military bases and many publications are available to researchers, making this area one of the best for research. The master description of government property throughout the world, the compilation of separate reports submitted by each government agency (including each of the military services) is *Summary Report on Real Property Owned by the United States Throughout the World, as of September 30, 19__* (GSA, Wash., D.C.: GPO). A description of property leased by the United States is *Summary Report on Real Property Leased to the United States Throughout the World as of September 30, 19__* (GSA, Wash., D.C.: GPO). The Army, Navy, and Air Force master real property listings provide complete information on occupancy and real property facilities value for military bases throughout the world:

Inventory of Air Force Real Property, 30 Sept. 19__ (Wash., D.C. (Bolling AFB): USAF, Dir. of Engineering and Services)

Inventory of Air Force Real Property, Overseas, 30 Sept. 19__ (Wash., D.C. (Bolling AFB): USAF, Dir. of Engineering and Services)

Inventory of Army Military Real Property, The United States, 30 Sept. 19__ (Wash., D.C.: USA, Office of the Chief of Engineers)

Inventory of Army Military Real Property, Outside the United States, 30 Sept. 19 __(Wash., D.C.: USA, Office of the Chief of Engineers)

Inventory of Military Real Property, Navy, 30 Sept 19__ (Alexandria, VA: USN, Naval Facilities Engineering Command)

The Department of Defense annual summary of property inventory and costs is *Real and Personal Property of the DOD* (Wash., D.C.: WHS (DIOR)). The Defense Department submits a quarterly report of property transactions to Congress—*Quarterly Report on the Acquisition and Disposal of Real Property* (Wash., D.C.: DOD, OASD (MRAL))—which updates the inventory.

Listings of military installations and information on military bases can also be obtained from other government publications. The *Base Structure Annex to the Manpower Requirements Report for FY__* (DOD, OASD(MRAL), Wash., D.C.: NTIS, 1976-) is an annual report and listing of military installations which makes the connection between the military force structure and the major military installations. It contains information on acreage and personnel and reports the status of the base structure and changes made or anticipated. Another publication which provides data on military installations and gives much descriptive data on facilities and capacity is the *Domestic Base Factors Report* (DOD, OASD (MRAL), Wash., D.C.: NTIS).

Pure listings of installations available are "Principal Military Installations or Activities in the 50 States" (OASD(PA) Factsheet, Processed), *U.S. Army Installations and Major Activities* (DA Pam 210-1), *USAF Installations Directory (Worldwide)* (AFP 87-13), *Standard Naval Distribution List, Parts 1 and 2*, and *List of Marine Corps Activities* (MCO P-5400.6). Other listings with descriptive information are "Air Force and Air National Guard Bases" in *Air Force Magazine* Almanac Issue (annually in May), which gives some information on size and personnel. This listing is derived from the Air Force Fact Sheet, "The USAF Base Guide" (AF Fact Sheet 78-5) distributed by Air Force Public Affairs. The other public affairs offices of the services also make available listings of military installations.

Two other Defense Department publications also of interest are *Map Book: Major Military Installations* (Wash., D.C.: DOD, WHS (DIO&R), 1976?-), a biennial publication of state maps showing major military installations and government owned and operated industrial facilities and *Distribution of Personnel by State: By Selected Locations, as of September 30, 19__* (Wash., D.C.: DOD, WHS(DIO&R), 1976?-), an annual publication of data on military personnel by state, area and location.

Military construction on bases influences continued and future occupancy, increases in personnel, and local employment and contracting. Information on military construction is contained in the comprehensive hearings held every year before the House and Senate Armed Services and Appropriations Committees and published in four sets of hearings transcripts. These hearings and the military construction portion of the budget breaks down construction projects by installation, explains their purpose and justifies them in terms of need and expenditures. The justification for construction and the overall description of the construction program usually includes a description of the base structure in general and a description of the bases where military construction is being requested. The breakdown of military construction projects by installation is contained in "Fiscal Year 19__ Military Construction Program" (OASD(PA) News Release), an annual breakdown. Each military construction project is described in detail in the justification material prepared for each appropriations title (see Table 23). The justification material is reproduced in the House Defense Appropriations Subcommittee hearings. The Defense Department also submits a number of reports to the Congress in the course of the year which reports on cost, design, obligations, overruns, etc. of the military construction program. The final amount by installation and project authorized by Congress is contained in *Enacted Summary of the New Construction Authority for Fiscal Year 19__* (Wash., D.C.: DOD, OASD(MRAL)).

The closing of military bases also affects the community (although in most cases new industries have made use of abandoned airstrips and other facilities). The federal government has set up an Economic Adjustment Committee (within the Department of Defense) which reports on economic adjustment programs at closed or reduced installations. The Economic Adjustment Committee studies the nature and the extent of damage or change caused by DOD vacation of facilities. Economic Adjustment Reports analyze an affected area as a result of a basing decision. The Defense Department's findings under the economic adjustment program have been optimistic and are summarized in *Summary of Completed Base Economic Adjustment Projects: 1961-1979* (Wash., D.C.: DOD, Office of Economic Adjustment, 1979). The American Enterprise Institute for Public Policy Research has also published a useful report on economic adjustment, *Military Base Closings: Benefits for Community Adjustment* (Wash., D.C.: AEI, 1977).

Good news about economic adjustment programs would lead one to relate these programs to conversion programs: the eventual (or immediate) conversion of military-related facilities into civilian-related facilities as a result of arms control and disarmament. Hearings, however, on Department of Defense base realignment policy (by the Senate Armed Services Committee in September, 1977, and August, 1978) indicate the importance of political decisions in policy formulation and implementation. One doing research on the local impact of military presence and spending should, therefore, follow DOD announcements of proposed base realignments and closures (for military utility or savings) and the actual number of implemented proposals.

G. OVERSEAS BASES, COMMITMENTS AND AID

An integral part of U.S. foreign policy and world influence

and exposure is the large number of overseas military bases and personnel, treaties and commitments, and military and security related assistance programs. The United States leads in every one of these areas and has continued to maintain large numbers of military personnel overseas and pursue an aggressive military aid and sales program to gain influence throughout the post-World War II period. Most aspects of military relations with foreign countries are well-documented. One cannot consider U.S. relations with foreign countries without considering U.S. military relations. This section is primarily concerned with overseas commitments, aid and sales and the presence of U.S. military forces overseas. It does not address the major U.S. alliance programs (NATO and Korea) which contribute to a large extent to U.S. overseas presence (they are addressed in Sections VD and VB3). It does, however, include general reference and descriptive works which include references to these programs.

1. *Commitments:* The central element of U.S. commitments to other countries are the treaties and agreements which commit the United States to come to the aid of another country, establish a relationship which leads to military aid and sales, or codify the presence of U.S. military forces in a foreign country. The annual annotated listing of these agreements is *Treaties in Force: A List of Treaties and Other International Agreements of the United States in Force as of January 1, 19__* (Dept. of State, Wash., D.C.: GPO, 1929-1956-). New treaties of the United States are produced as a pamphlet series, also prepared by the State Department—*Treaties and Other International Acts Series* (TIAS) (Dept. of State, Wash., D.C.: GPO, 1946)—and available by subscription from the GPO. Current treaty signings are announced in the *Department of State Bulletin.* Treaties are concluded for base agreements, military missions and weapons production. The full text of all current U.S. treaties is contained in *U.S. Treaties and Other International Agreements* (UST) (Dept. of State, Wash., D.C.: GPO, 1952-) which contains all of the agreements of the United States which have entered into force since 1950 (the new treaties are published in pamphlet form in the TIAS series until a new volume of the UST is printed).

An examination of the foreign relations of the United States is held every year as part of both the Defense Department and State Department hearings on the Defense and Foreign Aid and Security Assistance budgets. Periodic hearings are held by the House and Senate Foreign Affairs and Foreign Relations Committees on arms sales and economic and security assistance programs vis-a-vis particular countries and regions. The annual House Foreign Affairs foreign assistance hearings are a comprehensive record of U.S. relations throughout the world. These hearings normally cover the following areas:

Economic, Development and Military Assistance
Security Supporting Assistance
Africa
Human Rights
Europe and the Middle East
Asia
Latin America

The most comprehensive examination of U.S. commitments and aid was a series of hearings held in 1970 by the Senate Foreign Relations Committee. These hearings are still the source of a great deal of useful information:

U.S. Security Agreements and Commitments Abroad (Hearings before the Subcommittee on U.S. Security Agreements and Commitments of the Senate Committee on Foreign Relations, 91-2, Wash., D.C.: GPO, 1970)
Part 1: Philippines
Part 2: Kingdom of Laos
Part 3: Kingdom of Thailand
Part 4: Republic of China
Part 5: Japan and Okinawa
Part 6: Republic of Korea
Part 7: Greece and Turkey
Part 8: Ethiopia
Part 9: Morocco and Libya
Part 10: U.S. Forces in Europe
Part 11: Spain and Portugal

A summary of the hearings is *American Military Commitments Abroad* (Roland A. Paul, New Brunswick, NJ: Rutgers Univ. Press, 1973), by the counsel to the committee.

Another study conducted at the same time and still valuable for research is *Collective Defense Treaties* (Committee Print, House Committee on International Relations, 91-1, Wash., D.C.: GPO, 1969), a compilation of the provisions of the collective security treaties to which the United States is a party. More recently, numerous hearings have been held on certain aspects of U.S. policy in regions of the world and many of these are referenced in Section IVG3 below. One particularly useful study is a recent CRS report, *United States Foreign Policy Objectives and Overseas Military Installations* (Report of the Committee on Foreign Relations, U.S. Senate, by the Foreign Affairs and National Defense Division, CRS, LC, Wash., D.C.: GPO, 1979). This report describes U.S. bases throughout the world and their overall role in U.S. foreign and military objectives.

2. *Overseas Bases:* Central to U.S. overseas presence and influence are the military installations which the United States maintains in foreign countries. Most of these bases are in NATO countries and Korea and Japan, in support of U.S. commitments under bilateral and multilateral security agreements. As mentioned above, *United States Foreign Policy Objectives and Overseas Military Installations* is the best single study on the location and purposes of overseas bases. Another study of the CRS is *United States Military Installations and Objectives in the Mediterranean* (Report prepared for the Subcommittee on Europe and the Middle East of the Committee on International Relations, U.S. House, by the Foreign Affairs and National Defense Division, CRS, LC, Wash., D.C.: GPO, 1977), presenting the same type of material for U.S. installations in the Mediterranean region.

Many of the reports on military bases mentioned in the previous section also include information on overseas bases including listings and descriptions. The DOD inventories are the best source on all overseas bases, major and minor. As many country studies have shown, however, even the listing of military property is incomplete. Many overseas bases are secret, or belong officially to the host country. The *Base Structure Annex* gives much information on overseas bases including personnel and acreage. Overseas military installations are also identified by their distinctive Army Post Office (APO) or Fleet Post Office (FPO) numbers which are broken down in the Post Office *Zip Code Directory* and in AFM 10-5 and the *Standard Naval Distribution List, Part 1.* The Defense Department issues a quarterly News Release which presents the breakdown of U.S. military personnel throughout the world, by country and region: "Department of Defense Active Duty Military Personnel Strengths by Regional Area and by Country" (aka "U.S. Military Strengths Worldwide") (OASD(PA) News Release).

Other specific sources on overseas military installations are *U.S. Navy and Marine Corps Activities Outside the United States* (Dept. of the Navy, CHINFO Factsheet, n.d.), *U.S. Army Installations and Major Activities* and *USAF Installations*

Directory (Worldwide). Some studies of specific bases in foreign countries, many of which are politically sensitive, are mentioned in the next section. One general report which calls for American bases overseas as an instrument of foreign policy and power projection is *U.S. Overseas Bases: Problems of Projecting American Military Power Abroad* (Alvin J. Cottrell and Thomas H. Moorer, The Washington Papers #47 (CSIS), Beverley Hills, CA: SAGE, 1977).

3. Regional Presence and Relations: Sources on regions of the world, including the United States' presence and relations with foreign countries, is discussed in Section VB. All of the sources discussed here and other sources on these areas are covered more fully in that section.

The majority of U.S. forces and military bases are located in the NATO nations and support the U.S. commitment to NATO. Section VD deals specifically with NATO and includes many studies and sources which relate to the U.S. commitment and presence in that region. Some studies of value are *U.S. Naval Forces in Europe, U.S. Troops in Europe: Issues, Costs and Choices, United States Military Installations and Objectives in the Mediterranean* and the Congressional Budget Office series on *U.S. Air and Ground Conventional Forces for NATO*.

After Europe and NATO, the largest contingent of U.S. overseas military personnel and bases is in Asia and the Pacific. Some sources which provide information on U.S. military presence in that region are:

A Suitable Piece of Real Estate: American Installations in Australia (see Section VB2c)

Strategic Implications of American Bases in Australia (see Section VB2c)

Our Commitments in Asia (Hearings before the Subcommittee on East Asian and Pacific Affairs, House Committee on Foreign Affairs, 93-2, Wash., D.C.: GPO, 1974)

Planning U.S. General Purpose Forces: Forces Related to Asia (CBO, Wash., D.C.: GPO, June 1977)

Philippines Bases (CRS (Issue Brief 79041), Wash., D.C.: CRS, 1979)

The Military Equation in Northeast Asia (see Section VB2)

United States-Japan Security Relationship: The Key to East Asian Security and Stability (see Section VB2e)

The Indian Ocean in International Politics: Soviet-U.S. Rivalry and the Balance of Power: A Selected Bibliography (see Section VB).

Numerous sources relate to U.S. forces in Korea and they are discussed in Section VB2.

Sources on American security relationships with African nations are also discussed in Section VB. Some sources of interest are:

U.S. Military Involvement in Southern Africa
The U.S. and Africa: Guide to U.S. Official Documents and Government-sponsored Publications on Africa, 1785-1975
U.S. Security Interests and Africa South of the Sahara
A Survey of U.S. Government Investments in Africa
Options for U.S. Policy Toward Africa

Much less has been written of reference value specifically on U.S. military relations with the Middle East and Latin American regions. Again some sources of interest include:

American Policy Options in Iran and the Persian Gulf (see Section VB5d)
The Clouded Lens: Persian Gulf Security and U.S. Policy (see Section VB5d)
U.S. Intervention in Latin America: A Selected and Annotated Bibliography (see Section VB4)
Dimensions of U.S.-Latin American Military Relations (see Section VB4)

4. Aid and Military Assistance: U.S. military aid, sales and assistance form a portion of U.S. foreign policy which most visibly ties the United States to another country. Information on U.S. military aid and sales programs is quite extensive, although a great deal of the material requires compilation and analysis. Under the Arms Export Control Act of 1976 and other legislation, much information is made available to the Congress about U.S. military export programs, and thus much information is available from the Defense and State Departments. Besides official government sources, many military magazines report announcements of U.S. military sales and exchanges of military equipment and a number of research organizations monitor arms sales.

(a) Policy and Implementation: U.S. military sales and assistance to foreign countries are governed by the Arms Export Control Act, Foreign Assistance Act, and other legislation. The Carter Administration also adopted an arms sales policy which set certain restrictions on the conduct of the arms sales program

and the criteria and volume of sales and aid. These restraints, announced by President Carter on May 19, 1977, are described in *Arms Transfer Policy* (Presidential Report to Congress, released by the Subcommittee on Foreign Assistance, Senate Foreign Relations Committee, 95-1, Wash., D.C.: GPO, 1977). Two analyses of this policy are *Conventional Arms Transfer Policy: Background Information* (Prepared by the Subcommittee on International Security and Scientific Affairs, House International Relations Committee, 95-1, Wash., D.C.: GPO, 1978), and *Implications of President Carter's Conventional Arms Transfer Policy* (Report by the CRS for the Subcommittee on Foreign Assistance of the Senate Foreign Relations Committee, 95-1, Wash., D.C.: GPO, 1977). An integral part of the overall Carter Administration arms sales policy was its human rights provisions which are examined in *Human Rights: U.S. Foreign Policy* (CRS (Issue Brief 79056), Wash., D.C.: 1979), and the annual report of the State Department examining human rights throughout the world, *Report on Human Rights Practices in Countries Receiving U.S. Aid* (Report submitted to the U.S. Senate Foreign Relations Committee and the House Foreign Affairs Committee, by the Dept. of State, Wash., D.C.: GPO, 1978-).

U.S. arms sales policy in general is examined in two additional Congressional documents: *United States Arms Transfer and Security Assistance Programs* (Prepared for the Subcommittee on Europe and the Middle East, House International Relations Committee, by the CRS, LC, Wash., D.C.: GPO, 1978), and *Arms Sales: U.S. Policy* (CRS, (Issue Brief 77097), Wash., D.C., 1977). Much has been written on U.S. arms sales and military assistance policy and programs and much of this material is referenced in an excellent bibliography: *Foreign Military Sales of the United States: Selected References* (Air University Library: Maxwell AFB, AL: Aug. 1977 (Supplement 1, Nov. 1978)).

Three other studies worth mentioning are Congressional Budget Office studies on the effects of arms sales on the economy and the cost of weapons systems:

The Effects of Foreign Military Sales on the U.S. Economy (July 1976)
Budgetary Cost Savings to the Department of Defense Resulting from Foreign Military Sales (May 1976)
Foreign Military Sales and U.S. Weapons Costs (May 1976)

Internal governmental implementation of arms sales policy is described in the *Military Assistance Sales Manual* (DOD

5105.38M) (Wash., D.C.: DOD DSAA, 1 Aug 1978), which outlines the responsibilities, policies, and procedures governing the administration of security assistance programs of the DOD, including information on the planning, execution and purposes of the various programs. Each of the military services have implementing and supplemental logistic and policy regulations dealing with security assistance and these should also be consulted for advanced research dealing with the implementation and administration of military sales programs. An industry organization, the American Defense Preparedness Association (ADPA), covers much of the same ground in *How to Conduct Foreign Military Sales* (William H. Cullin, Arlington, VA: ADPA, 1977).

Three newsletters within the military assistance community dealing with policy are also of value:

DISAM Newsletter (Defense Institute for Security Assistance Management, Wright Patterson AFB, OH), which covers security assistance policy and administration, with a directory of DOD offices involved in security assistance policy, administration and implementation.

U.S. Army Security Assistance Bulletin (Wash., D.C.: DA, ODCSOPS, 1973-), a semiannual newsletter of policy announcements for the Army.

U.S. Navy Security Assistance Newsletter (Wash., D.C.: DN, OPNAV (OP-63)).

(b) Current Programs: Current military assistance and sales programs are outlined every year as part of the Congressional hearings on the Foreign Assistance program. The hearings usually cover all of the general and policy aspects of U.S. security assistance programs. The Department of State and the Department of Defense develop justification material for the Security Assistance programs, which should be consulted for the most exact information on administration requests. The *Congressional Presentation Document, Security Assistance Programs, FY__* (Department of State, Wash., D.C.) contains information on military aid programs, Foreign Military Sales, the Military Assistance Program, International Military Education and Training (IMET) programs, and Security Supporting Assistance, by country. The statistics on actual sales and aid are contained in the annual *Foreign Military Sales and Military Assistance Facts* (Wash., D.C.: DSAA, Data Management Division, Dec. 19__), which contains statistics on all of the above areas and includes commercial exports, agreements, deliveries and financing and costs by country and by year. *U.S. Overseas Loans and Grants and Assistance to International Organizations: Obligations and Loan Authorizations* (AID, Wash., D.C.: GPO, 1971-) presents the value of all economic and military assistance tabulated by country and itemized by program from 1945 to the present. Annual tabulations of arms exports through the Commercial Sales Program are published by the State Department's Office of Munitions Control (OMC) as the *Report Required by Section 657, Foreign Assistance Act* (Wash. D.C.: OMC).

Data of specific transactions of weapons systems is compiled by the SIPRI and made available in the *SIPRI Yearbook* annually. *The Military Balance* also contains data on major arms transactions of the previous year. *Arms Trade Data* (Daniel Volman and Michael Klare, comps., Wash. D.C.: Institute for Policy Studies, 1978 (revised 1981)) is a comprehensive reference source on U.S. military sales to Third World countries from 1973 to the present. U.S. arms sales to police forces of Third World countries are tabulated in *Supplying Repression* (2nd Ed.) (Cynthia Arnson and Michael T. Klare, Wash., D.C.: Institute for Policy Studies, 1981). Statistics on the international arms traffic are contained in the ACDA's *World Military Expenditures and Arms Transfers* (Wash., D.C.: ACDA), published annually with data for the preceding ten years. A CIA study, *Arms Flow to LDCs: U.S. Soviet Comparisons, 1974-77* (Wash., D.C.: CIA (ER 78-10494U), 1978), is also worth mention, as is the annual CIA survey of Communist economic and military aid and assistance programs: *Communist Aid Activities in Non-Communist Less Developed Countries, 19__* (Wash., D.C.: CIA, 1976?-).

Finally, there are two sources which try to identify U.S. products and the market for U.S. arms sales. *Arsenal of Democracy: American Weapons Available for Export* (Tom Gervasi, New York: Grove Press, 1977), a directory of American weapons, presents information on U.S. military sales programs and policy.

Foreign Military Markets (DMS Market Intelligence Reports) (Greenwich, CT: DMS, Inc.) is an information service of a base volume with monthly updates of reports on the military organization, posture and budget of 86 countries that are potential U.S. military arms sales markets with information on arms needs, procurement organization, current suppliers and manufacturing capability.

V: WORLDWIDE MILITARY AND STRATEGIC AFFAIRS

The sources introduced in the first four parts of this guide have dealt with progressively more specific issues of military and strategic affairs. The previous part dealing with the U.S. military presented sources which were limited for the most part to specific aspects of American military force, policy and data. But no meaningful research on military affairs can be conducted without considering the role of other countries. This final part presents sources of information dealing with foreign countries and with transnational military issues. Sources have been chosen to provide sufficient background to conduct both raw primary and secondary research. Information services are discussed in the first section, and sources of information on specific countries of the world are then presented. Numerous subject areas transcending all nations are also discussed: weapons, arms control, and international organizations.

As in other parts of this guide, sources have been limited to relatively accessible English language works which have proven valuable to researchers. Many are well-known reference books and some are more recondite and specialized. An attempt has been made to present primarily recurring publications, bibliographies and periodicals.

A. ARMED FORCES AND WEAPONS

In Section IIB, general sources on quantitative aspects of military forces were presented as an introduction to sources on military forces worldwide. This section presents more specific information sources on military forces and weapons and also presents the periodicals and other sources that provide the raw material for military analysis and trends and current events research. Like any other field, there is an abundance of materials directed at many different audiences. Most of the sources chosen for reference in this part are information services, periodicals and reference works that are directed at defense analysts in government, industry and the media. It presents both introductory and the more sophisticated material.

1. Information Services: A number of information services and information houses or brokers produce a great deal of the material available on military forces worldwide and weapons and weapons programs. These are included together in this part because most of these services provide very specific data on military forces and weapons of many countries but are not for the most part introductory sources. The products of the most well-known research organizations have already been mentioned. The London International Institute for Strategic Studies' (IISS) products are the most accessible and authoritative introductory sources. *The Military Balance, Strategic Survey,* and *Adelphi Papers* from IISS receive wide distribution and are thus worth mentioning again. Many of the other information and research organizations and their products and information services are more obscure and expensive, and therefore less accessible.

The largest information service in the military forces and weapons field is Defense Marketing Service, Inc. (100 Northfield St., Greenwich, CT 06830), which produces a variety of information products, newsletters and special reports. Most of DMS's products are very expensive and are directed at defense industries. Table 25 is a breakdown of the various products. The newsletters are informative and geared toward the possible military sales or contracting markets open to defense producers. The *Market Intelligence Reports, Market Studies* and *Special Reports* present data on the products, producers, budget, and defense programs in the specific military fields. The *Foreign Military Market Reports* describe the organization of military forces of 86 countries, with information geared towards foreign military sales. The *Aerospace Agencies, Aerospace Companies, "AN" Equipment, Defense Market, Code Name Handbook,* and *Congressional Testimony Index* are unique and valuable reference works, specifically for budget and program analysis.

Another large information broker which produces solely market research reports (for defense industry) is Frost & Sullivan, Inc. (106 Fulton St., New York, NY 10038). Frost & Sullivan contracts to have large market analyses done on specific defense product areas which normally include an analysis and forecast of opportunities in the field, current defense programs, manufacturers, and research and development projects. These reports cost around $800 dollars per study. The titles of some recent reports are:

TABLE 25
INFORMATION SERVICES AND PRODUCTS OF DMS, INC.

Market Intelligence Reports (base reference volumes with monthly updates):

Aerospace Agencies	Military Aircraft
Aerospace Companies	Missiles/Spacecraft
"AN" Equipment	Ships/Vehicles/Ordnance
Civil Aircraft	Commercial Laser & EO
Defense Market	[electro-optics]
Electronic Systems	Military Laser and EO
Gas Turbine Engines/	Electronic Warfare
Gas Turbine Market	Radar and Sonar
	Military Simulators

Foreign Military Markets (base volume and monthly updates):

NATO/Europe	South America/Australasia
Middle East/Africa	NATO Weapons

International Market Intelligence Reports (base volume and monthly updates):

Military Communications (Europe)	Missiles and Satellites (Europe)
	Warships (Europe)

Market Studies and Special Reports (annual):

Code Name Handbook	U.S. Government Civil
Congressional Testimony Index	Laser and EO
Defense RDT&E Budget Handbook	Simulators—Market Study and Forecast
Defense Procurement Budget Handbook	World Aircraft Forecast to 1988
Foreign Military Laser and EO	World Helicopter Forecast to 1988
	Operations and Maintenance

Defense Research and Development Series (R&D program analysis and forecast):

Command/Control/Communications	Fire Control Systems
Computers	Missiles and Rockets
	Navigation/Guidance

Newsletters:

Aerospace Intelligence	International Defense Intelligence
Contracting Intelligence	Turbine Intelligence

Specially tailored computer-based contracting services on Company, Agency and Contracting information.

The ASW [Anti-Submarine Warfare] *Market*
The Electronic Warfare Market
The Military Airborne Radar Market
The Military Computer Market
The Military Communications Market
The Military Satellite Communications Market
The Military Reconnaissance and Surveillance Market
Military Surface-Launched Tactical Missiles & Support Equipment

Frost & Sullivan also sells computer-generated reports covering contracting information retrievable by program, contractor or agency. These tailored information services are especially valuable for advanced research on the military-industrial complex.

Other organizations more competitive with the International Institute for Strategic Studies in terms of information on military forces and military developments are Aviation Studies Atlantic, Aviation Advisory Services, Sorecom SAM/Interinfo, and Interavia. Aviation Studies Atlantic (Sussex House, Parkside, London SW19, U.K.) produces loose-leaf information services in many areas. These services, which also include a base volume and periodic updates, are very specific and sometimes extremely difficult to use due to poor organization. Some of the information services include:

Army, Air Force and Naval Air Statistical Record
Army Vehicles and Military Aircraft Data Sheets
Forecast Data Bank Cumulative Sheets
Armaments Data Sheets
Official Price Lists
Nuclear Weapons Data File
Military Record of Atomic/CBR [Chemical, Biological and Radiological] *Happenings*

Some of these services, specifically the *Nuclear Weapons Data File* and the *Official Price Lists* are unique and quite valuable. In general, however, Aviation Studies Atlantic material is poorly organized and difficult to use.

Aviation Studies Atlantic's biggest competitor is Aviation Advisory Services which produces the *International Air Forces and Military Aircraft Directory*. This reputable and well-organized loose-leaf service covers air matters worldwide and is updated by the valuable and authoritative monthly newsletter, *Milavnews*.

Interavia, SA (86 Avenue Louis-Casai, 1216 Cointrin, Geneva, Switzerland) and Sorecom SAM/Interinfo (Rue des Orchidees, Monte Carlo, Monaco) are prestigious magazine publishers in the international defense fields which have information services. Interavia, publisher of the excellent *International Defense Review,* also publishes the *Interavia Air Letter,* and has a number of loose-leaf services and surveys dealing with military weapons and forces. Some of the Interavia Data products are:

Air Forces of the World
Aircraft Armament
Military Avionic Equipment
Military Communications
Aircraft Production Survey
Current Aircraft Prices

Sorecom publishes *Ground Defence International* and *Aviation and Marine International* (as well as a number of French and Arabic language military journals) and has some information services, mostly for custom research for a fee.

Finally there are the newsletters, mostly from the Washington area, which report on military or contracting activity and are directed at the defense industries, lobbyists and analysts. Table 26 lists the newsletters—most of which run over $100 per year—and the organizations which produce them. These are limited to a few large publishing houses.

Copley & Associates, SA (Suite 602, 2030 M St., N.W., Washington, D.C. 20036) and Callahan Publications (6631 Old Dominion Drive, McLean, VA 22101) are the largest newsletter publishers, publishing five and four newsletters respectively. Government Business Worldwide Reports (P.O. Box 5651, Washington, D.C. 20016) publishes newsletters but also has an information service offering various publications on budgets, defense programs, and research and development.

2. Weapons Series: In addition to the information services and newsletters mentioned above, there are a number of standard sources which include information on weapons systems, characteristics, and trends in research and development, production and capabilities. The *Market Intelligence Reports* of DMS, Inc., and the *Market Reports* of Frost & Sullivan are excellent sources of program information, but the weapons series of Jane's, Brassey's and the "Of the World" series, as well as numerous government reference works, are more specific and

TABLE 26
MILITARY NEWSLETTERS

Aerospace Daily (Ziff-Davis)
Aerospace Intelligence (DMS)
Contracting Intelligence (DMS)
Defense & Economy World Report & Survey (Government Business Worldwide Reports)
Defense & Foreign Affairs Daily (Copley)
Defense Business/International Defense Business (Government Business Worldwide Reports)
Defense Daily (Space Publications)
Defense Week
The Government Contractors Communique (Federal Publications)
Interavia Air Letter (Interavia)
International Defense Intelligence (DMS)
Milavnews (Aviation Advisory Services)
Military Research Letter (Callahan)
Missile/Ordnance Letter (Callahan)
Renegotiation/Management Letter (Callahan)
Shield Newsletter (Whitton Press)
Soviet Aerospace (Space Publications)
Strategic Latin American Affairs (Copley)
Strategy Week (Copley)
Turbine Intelligence (DMS)
Underwater Letter (Callahan)

are indispensible.

Macdonald and Jane's (Paulton House, 8 Shepherdess Walk, London N1, U.K.) (U.S. distributor: Franklin Watts, Inc., 730 Fifth Ave., New York, NY 10019) is the largest and most prestigious and authoritative publisher of weapons reference books with a series of major reference books and pocket books that are foremost in their field. The major reference books are approaching prohibitive costs (circa $100 per volume) but are available in most libraries. There are ten military titles in the weapons systems yearbook series and 23 volumes in the pocket book series (see Table 27). The titles in the weapons systems yearbook series are:

Jane's Armour and Artillery
Jane's All the World's Aircraft
Jane's Fighting Ships
Jane's Infantry Weapons

Jane's Weapons Systems
Jane's Freight Containers
Jane's Military Vehicles and Ground Support Equipment
Jane's Surface Skimmers
Jane's Ocean Technology
Jane's Military Communications

Brassey's publishes a number of reference works on military equipment also of value. The *RUSI/Brassey Defence Yearbook* has a weapons review every year which is also published as *International Weapons Developments* (London: Brassey's/San Rafael, CA: Presidio Press, 1979). This reference book covers the development and status of major Army, Navy, and Air Force weapons, electronic warfare, communications equipment, and future trends. The Brassey's reference books on weapons systems deal mainly with ground systems and include:

TABLE 27
JANE'S POCKET BOOKS

Book 1	Major Warships
Book 2	Major Combat Aircraft
Book 3	Commercial Transport Aircraft
Book 4	Modern Tanks and Armoured Fighting Vehicles
Book 5	Military Transport and Training Aircraft
Book 6	Light Aircraft
Book 7	Airship Development
Book 8	Submarine Development
Book 9	Naval Armament
Book 10	Missiles
Book 11	Space Exploration
Book 12	Research and Development Aircraft
Book 13	RPVs: Robot Aircraft Today
Book 14	Home Built Aircraft
Book 15	Record Breaking Aircraft
Book 16	Pistols and Sub-machine Guns
Book 17	Rifles and Light Machine Guns
Book 18	Towed Aircraft
Book 19	Heavy Automatic Weapons
Book 20	Helicopters
Book 21	Hovercraft and Hydrofoils
Book 23	Electrical Locomotives

Brassey's Infantry Weapons of the World, 1950-1975: Infantry Weapons and Combat Aids in Current Use by the Regular and Reserve Forces of the World
Brassey's Artillery of the World
Brassey's Fast Attack Craft

The "Of the World" series of weapons reference books are cheaper, more general reference books. Many of these publications are edited by Christopher Foss and published in London but also republished in the United States. The titles in the "Of the World" series are:

Armoured Fighting Vehicles of the World
Artillery of the World
Infantry Weapons of the World
Military Vehicles of the World
Warships of the World
Helicopters of the World
Missiles of the World

The government publishes some weapons reference books, which are often quite specialized and of great interest because of their specific nature and authority. The *U.S. Army-Europe Identification Guides to Soviet and East European Equipment* are valuable sources of unclassified information on specific Soviet ground systems. The titles in the series are:

Part One: Weapons and Equipment: East European Communist Armies
Volume I: General Ammunition and Infantry Weapons (30 Sept. 1972)
Volume II: Artillery (15 Jan. 1975)
Volume III: Armored Vehicles, Tanks and SP Artillery (15 Feb. 1973)
Volume IV: Armored Vehicles, Scout Cars, APCs, and Tank Recovery Vehicles (1 Mar. 1973)
Part Two: Weapons and Equipment: East European Communist Armies
Volume I: Tractors and Trucks, Amphibious Vehicles, Snow and Swamp Vehicles (10 Oct. 1973)
Volume II: Soviet Trucks and Trailers (15 Feb. 1974)
Volume III: Non-Soviet Trucks and Trailers (15 Jul. 1974)
Part Three: Weapons and Equipment: East European Communist Armies
Volume I: Bridging and Stream Crossing Equipment (25

Apr. 1975)

Volume II: Mine Warfare and Demolition Equipment (30 Jul. 1975)

The Army Foreign Science and Technology Center issues a series of technical intelligence publications, the *Foreign Materials Catalog,* many of which are unclassified. The unclassified titles in the series are:

Volume 1: Infantry Weapons
Volume 6: Fuzes
Volume 11: Aerial Delivery Equipment
Volume 12: Gap Crossing Equipment
Volume 13: General Engineer Equipment
Volume 14: Construction Equipment
Volume 15: Power Generating Equipment
Volume 16: Engines
Volume 17: General Equipment
Volume 19: Materials Handling Equipment
Volume 20: Military Subsistence
Volume 21: POL Handling Equipment
Volume 22B: Surface Transport Equipment (Free World)
Volume 22C: Special Purpose Vehicles

The Australian and British governments also publish weapons reference books which are directed at potential buyers of defense materiel. These sources, called *Defence Equipment Catalogues,* are excellent sources for information on equipment of British or Australian manufacture.

3. *Air Forces:* More information is available on air forces than on any other branch of the military. Interest in military aircraft and aviation is very high among defense analysts and enthusiasts, and a wide range of periodicals and reference books is available.

The various sources on air forces of the world range from specialized original information reference books and services to weapons directories with reproduced information taken from *The Military Balance* and aimed at the enthusiast. The best sources of original information on details of force structure, equipment and programs are:

International Air Forces and Military Aircraft Directory (Essex, U.K.: Aviation Advisory Services, Ltd., 1964-), the best source, a loose-leaf service with monthly updates covering organization, equipment and procurement plans of the air forces of the world. Updated monthly by the newsletter, *Milavnews.*

"World's Air Forces," Special Issue, *Flight International,* August 19__, annual country-by-country review of recent developments, with data and equipment descriptions.

Army, Air Force and Naval Air Statistical Record (London: Aviation Studies Atlantic, 1956-), a loose-leaf service with bimonthly updates covering the weapons inventories and order of battle of 104 countries.

Army Vehicle and Military Aircraft Data Sheets (London: Aviation Studies Atlantic, 1954-), a loose-leaf service with quarterly updates on the characteristics, number built, number on order and inventory of various weapons systems.

Air Forces of the World: An Illustrated Directory of All the World's Military Air Powers (Mark Hewish, et. al., New York: Simon and Schuster, 1979), organization, aircraft inventory, procurement, combat capability and bases of the air forces of 125 countries.

Air Forces of the World (Geneva: Interavia, SA), a loose-leaf service with quarterly updates on organization and weapons and programs of air forces of the world.

Other sources which deal with the air forces of the world and present information on weapons but which do not present as much data on the organization or plans of the forces are:

World Combat Aircraft Directory (Norman Polmar, ed., Garden City, N.Y.: Doubleday, 1976), worldwide brief air order of battle and descriptions (with photos) of about 200 of the major combat aircraft of the world.

World Military Aviation: Aircraft, Air Forces, Weaponry and Insignia (Nikolaus Krivinyi, et al., New York: Arco, 1977), brief descriptions of the organization and equipment (with line drawings), by country, with descriptions of weapons characteristics.

"Aerospace Forecast and Inventory Issue," special March issue, *Aviation Week and Space Technology,* an annual review of programs and trends in military, space, avionics programs worldwide with numerous "specification" tables with data on international aircraft, missile, and weapons programs and systems.

The *Forecast Data Bank Cumulative Sheets* (London: Aviation Studies Atlantic, 1956-) and the *Forecast Data Bank Cumulative Sheets Supplement* (London: Aviation Studies Atlantic, 1956-) also provide valuable reference information on 10-year forecasts of the inventory and procurement of 148 countries. These two loose-leaf services provide information by weapon system and country on civil and military aircraft, missiles, armored vehicles and other pieces of military equipment.

All of the above sources contain information on air forces or air force weapons. A number of periodicals also report developments relating to the force structure or capabilities of the world's air forces. These are listed in Table 28.

(a) Aircraft and Weapons: Most of the sources listed above have also dealt with aircraft and weapons in their description and discussion of air forces. Some sources which present information only on aircraft and air-related weaponry are also valuable. *Jane's All the World's Aircraft* (London, Macdonald and Jane's, 1919-1945-), issued annually, is the most reputable reference work on aircraft characteristics. Other sources which deal with aircraft include:

Aircraft Armament (Geneva: Interavia, SA), a loose-leaf service with semiannual updates dealing with developments in aircraft armaments.

Military Avionic Equipment (Geneva: Interavia, SA), a loose-leaf service with semiannual updates dealing with aircraft

---

TABLE 28
PERIODICALS DEALING WITH AIR FORCES AND WEAPONS

Aeronautical Quarterly	*Defence Materiel*
Aerospace (U.K.)	*Defence Today*
Aerospace International	*Flight International*
Air Combat	*Helicopter*
Air Force Magazine	*Interavia*
Air International	*International Defense Review*
Air Pictorial	*Jane's Defence Review*
Aviation & Marine International	*Journal of Aircraft*
Aviation Week and Space Technology	*Military Technology and Economics*
Canadian Aeronautics and Space Journal	*National Defense*
Defence	*NATO's Fifteen Nations*
	Naval Forces

electronics.

Aviation Reports (London: Aviation Studies Atlantic, 1951-), semiweekly reports with quarterly supplements of aircraft developments.

Aviation Studies Atlantic Official Price List (London: Aviation Studies Atlantic, 1954-), a loose-leaf service with quarterly supplements, of prices of aircraft throughout the world.

Helicopters of the World (Michael J.H. Taylor and John W.R. Taylor, New York: Scribner, 1978), a directory of helicopters in use worldwide.

"Military Aircraft Census," Special Issue, *Flight International,* September, an annual special issue of military aircraft inventories of countries of the world.

"Military Aircraft of the World," Special Issue, *Flight International,* March, an annual special issue of military aircraft characteristics and programs worldwide.

World Combat Aircraft Directory.

The World's Air Forces (Chris Chant, ed., Secaucus, NJ: Chartwell Books, 1979), an enthusiast's directory of equipment and characteristics.

The Observer's Soviet Aircraft Directory.

4. Ground Forces: Much less information is available on the organization and capabilities of the ground forces of the nations of the world than on the air forces. In fact, until recently, there was not even a general reference work which dealt with background information on the world's armies and ground forces. There is still only one authoritative reference source in this area: *World Armies* (John Keegan, New York: Facts on File, Inc., 1979). This is a country-by-country survey of the armies of almost every nation covering history, command and constitutional status, role, commitment, deployment and recent operations, organization, personnel recruitment and training, equipment and current developments. Two other sources present similar information for the armies of the nations of Europe and the Middle East: *The Armies of Europe Today* (Otto Von Pivka, Reading, U.K.: Osprey, 1974), and *Armies of the Middle East* (Otto Von Pivka, New York: Mayflower Books, 1979). These are country-by-country surveys of the organization, hierarchy, background, and equipment of the NATO, Warsaw Pact, neutral armies of Europe, and the militaries of the Middle East. The *Army, Air Force and Naval Statistical Record, Army Vehicle and Military Aircraft Data Sheets, International Air Forces & Military Aircraft Directory,* and *Forecast Data Bank Cumulative Sheets Supplement* also present information broken down by country on weapons inventories and structure of army aviation components and armies.

Periodicals dealing with the ground forces are of value since they contain information on exercises, weapons developments, structure and capabilities of the armies of the world. Table 29 lists some periodicals which cover army developments and weapons.

(a) Weapons: There are a number of reference works on ground forces equipment. Already mentioned are the government reference guides to ground forces equipment—the Army Foreign Science and Technology Center *Foreign Materials Catalog* and the *U.S. Army-Europe Identification Guides.* The Jane's books dealing with ground forces equipment—*Jane's Infantry Weapons* (London: Macdonald and Jane's/New York: Franklin Watts), *Jane's Weapons Systems* (London: Macdonald and Jane's/New York: Franklin Watts, 1969-), and *Jane's Combat Support Equipment* (London: Macdonald and Jane's/New York: Franklin Watts)—are the most authoritative for characteristics and developments. They are issued annually or biennially. The Brassey's reference books—*Brassey's Infantry Weapons of the World, 1950-1975: Infantry Weapons and Combat Aids in Current Use by the Regular and Reserve Forces of the World* (J.I.H. Owen, ed., New York: Crane, Russak) and *Brassey's Artillery of the World* (Brigadier R.G.S. Bidwell, ed., Boulder, CO: Westview, 1977)—are also reliable references for these weapons. The "Of the World" series dealing with ground forces equipment—*Armoured Fighting Vehicles of the World* (London: Allan/New York: Scribner, 1977), *Artillery of the World* (New York: Scribner, 1976), *Infantry Weapons of the*

TABLE 29
PERIODICALS DEALING WITH GROUND FORCES AND WEAPONS

Armada International	*Ground Defence International*
Army	*Helicopter*
Army Aviation	*International Defense Review*
Army Research, Development and Acquisition	*Jane's Defence Review*
	Military Technology and Economics
Defence	
Defence Materiel	*National Defense*
Defence Today	*NATO's Fifteen Nations*

World (New York: Scribner, 1977), *Military Vehicles of the World* (New York: Scribner, 1979), and *Helicopters of the World*—contain photographs and descriptions of major weapons in use and their variations and modifications.

Jane's has also published a separate reference book, *Jane's World Armoured Fighting Vehicles* (Christopher F. Foss, New York: St. Martins, 1976), which is as comprehensive and authoritative as its main series. The Defense Intelligence Agency (DIA) publishes a series of small arms identification and characteristics guides which are certainly authoritative: *Small Arms Identification and Operation Guide—Free World* (Wash., D.C.: DIA (DST-1100H-163-76), Dec. 1976), and *Small Arms Identification and Operations Guide—Eurasian Communist Countries* (Wash., D.C.: (DST-1110-H-394-76), Dec. 1976). *The World's Armies* (Chris Chant, Secaucus, NJ: Chartwell, 1979) highlights the most common ground pieces of military equipment including artillery, armored vehicles and air defense artillery but is directed primarily at enthusiasts and therefore limited in its value.

A number of valuable reference works deal specifically with the armies of certain countries or a region (e.g., the NATO and Warsaw Pact armies). These are discussed under the regional or country sections later in this guide.

5. Naval Forces: Although the quality of information on naval forces of the world varies greatly, some excellent sources are available to the analyst compiling a profile of the naval situation in a certain region of the world or a certain country. Since the relationship between naval force structure and equipment is very evident, the sources on naval ships and weapons provide a great deal of explanatory data on the characteristics and capabilities of the world's navies. Very few sources describe the organization and background of the navies. Most reference works break down naval capability in terms of numbers of types of ships. While much information is available on the U.S. and Soviet navies, their peacetime organization, training, background, and capabilities, much less has been compiled on the lesser navies. *Combat Fleets of the World, 19__-__: Their Ships, Aircraft and Armament* (Jean Labayle Couhat, ed., Annapolis, MD: USNI, 1978?-) is the best general reference book, covering the naval inventory, construction and merchant marines of 122 countries. *Guide to Far Eastern Navies* (Barry M. Blechman and Robert P. Berman, eds., Annapolis, MD: USNI, 1978) is another excellent source covering the background, organization, status, ships and aircraft of six Asian nations (China, Japan, North

Korea, South Korea, the Philippines, and Taiwan), and the naval balance in the Western Pacific. *Unclassified Communist Naval Order of Battle* (Wash., D.C.: DIA, May-Nov. 1974?-), is a semiannually updated, very authoritative source on force structure of communist countries.

Three sources give information on naval air components. The best is *International Air Forces and Military Aircraft Directory* which fully describes the naval aviation forces. The *Army, Air Force and Naval Air Statistical Record* and the *Forecast Data Bank Cumulative Sheets* describe naval air forces and naval aviation aircraft.

The periodicals which cover worldwide naval developments including profiles of individual country navies and ship types are listed on Table 30. These magazines report developments in naval ships, weapons and armament, aircraft and aviation, tactics and doctrine and forces and organization.

(a) Ships and Weapons: The sources above and those on general military forces discussed in Section IIB in many cases list navies by ship types and classes. The sources on naval ships and weapons provide the explanatory information describing the characteristics and capabilities of the various naval ships and classes. The best reference volume is again *Jane's Fighting Ships* (London: Macdonald and Jane's, 1898-1945-), an annual

---

TABLE 30
PERIODICALS DEALING WITH NAVAL FORCES AND SHIPS AND WEAPONS

Australian Naval Institute Journal
Aviation & Marine International
Aviation Week & Space Technology
Defence
Defence Materiel
Defence Today
Hovering Craft and Hydrofoil
Interavia
International Defense Review
Jane's Defence Review

Maritime Defence
Military Technology and Economics
National Defense
NATO's Fifteen Nations
Naval Forces
Naval Record
Navy International
Seapower
Sea Technology
Ships Monthly
USNI Proceedings
Warship
Zosen

147

reference work on the ships of the world. Other Jane's books which deal with naval ships and equipment include *Jane's Surface Skimmers* (New York: Franklin Watts) and *Jane's Ocean Technology* (New York: Franklin Watts). Other naval weapons and ships reference books of varying utility include *Brassey's Fast Attack Craft* (John Marriot, ed., New York: Crane, Russak, 1978) covering smaller naval ships, *Warships of the World* (Bernard Ireland, New York: Scribner, 1976), *Glossary of Naval Ship Types* (Wash., D.C.: DIA (DDB-1200-47-78), 1978), and *The World's Navies* (Chris Chant, ed., Secaucus, NJ: Chartwell, 1979). The periodicals in Table 30 are also sources on weapons and ships.

6. Nuclear Weapons: There is a lot of interest in nuclear weapons as a separate category but the paucity of information in the area of nuclear weapons inventories and forces makes original research in this area difficult. The only independent reference work which deals exclusively with nuclear weapons is the *Nuclear Weapons Data File: Order of Battle, Production, Development, Costs and Status of More Than 50,000 Nuclear Weapons in Six Countries Including New Dimensions* (London: Aviation Studies Atlantic). This loose-leaf service with periodic updates includes reference information on the order of battle (inventory and types), characteristics, technology, costs, testing and future trends of nuclear weaponry. Of course there is no way of confirming or judging the accuracy of the information in this work, but much of the data presented (yields, costs, exact inventories) is not in the public domain, making this a unique source. Another information service of Aviation Studies Atlantic is *Military Record of Atomic/CBR Happenings* (London: Aviation Studies Atlantic, 1954-), a loose-leaf service with quarterly supplements and updates of developments in the nuclear, chemical, biological and radiological warfare fields.

A great deal of technical and scientific material is written about nuclear technology (much of it related to peaceful uses of nuclear technology). This material is indexed in *ATOMINDEX* (Vienna, Austria: International Atomic Energy Agency), which incorporates the discontinued *Nuclear Science Abstracts* of the defunct Energy Research and Development Agency. Two bibliographies of nuclear-related material are *Nuclear Related Terrorist Bibliography* (Wash., D.C.: DA, Inscom (CID/DNA), 5 July 1979), an unannotated bibliography of works concerning nuclear-related terrorist incidents, scenarios and tactics; and *NEWS (Nuclear Energy, Weapons and Safeguards) Data Base: A Computerized Bibliography 1975-April 1978* (G. Petty, et al.,

Santa Monica, CA: Rand Corporation (P-6219), Oct. 1978), an unannotated bibliography of works covering the technical, political and institutional factors affecting nuclear weapons production and proliferation with an author, title and subject index. Other reference works on nuclear weapons and nuclear weapons proliferation are discussed in Section VE4.

The basic reference book on the effects of nuclear weapons is *The Effects of Nuclear Weapons* (3d Ed.) (Samuel Glasstone and Philip J. Dolan, comps., DOD, Defense Nuclear Agency, Wash., D.C.: GPO, 1977), which describes the radiological, blast, thermal, radio blackout and electromagnetic pulse effects of nuclear detonations. Another basic reference source, *The Effects of Nuclear War* (Office of Technology Assessment, Wash., D.C.: GPO, 1979) describes the effects of nuclear war in the United States. The *Physical Vulnerability Handbook: Nuclear Weapons* (Wash., D.C.: DIA, (AP-550-1-2-69-INT), June 1969) (with changes) is a basic source on hardening, vulnerability and destruction by nuclear weapons.

Other sources on nuclear weapons are the annual hearings of the Armed Services and Appropriations Committee on the "Atomic Energy Defense Activities" (nuclear weapons production and research) of the Department of Energy. The Department of Energy Public Affairs Office (1000 Independence Avenue, S.W., Wash., D.C. 20584) is a source of information on nuclear weapons programs, including information on nuclear weapons contracting. The Arms Control and Disarmament Agency can also provide information on nuclear weapons and the effects of nuclear war. It compiles the *Arms Control Impact Statements*, which contain much information on nuclear weapons production and programs.

Much alternative information on nuclear weapons and nuclear war is also available in magazines, from peace and arms control groups and from research institutes. Magazines such as the *Bulletin of the Atomic Scientists* and *The Progressive* have regular articles on nuclear weapons. Organizations such as SANE, NARMIC, the Nuclear Weapons Facilities Project of AFSC, or the Center for Defense Information specialize in nuclear weapons analysis. *The Counterforce Syndrome: A Guide to U.S. Nuclear Weapons and Strategic Doctrine* (Robert C. Aldridge, Wash., D.C.: Institute for Policy Studies, 1978) is an excellent summary of nuclear weapons developments.

7. Missiles: Of the specialized sources dealing with missile systems, *Jane's Weapons Systems* is very good and includes information on the characteristics and capabilities of the various systems. Other sources include:

U.S. Missile Data Book, 1978 (3d Ed.) (Ted G. Nicholas, Fountain Valley, CA: Data Search Associates, October 1978), includes information on the characteristics, production quantity, cost and inventory of U.S.-produced missiles.
Missiles of the World (Michael J.H. Taylor and John W.R. Taylor, New York: Scribners, 1976), a pocket book of photographs and descriptions.
The World's Missile Systems, 1977 (4th Ed.) (Pomona, CA: General Dynamics, 1977), a comprehensive reference work of the characteristics and developments in the world's missile systems.
"World Missile Directory," Special Issue, *Flight International*, June, annually, description and developments from antitank to strategic ballistic.

In addition, DMS, Inc. (see Table 25) has a number of information services which relate to missile systems.

8. Military Communications and Electronic Warfare: The growth of technology and the development of "smart weapons" and other advanced systems is a result of great advances in the communications and electronics field. The communications and electronics industry and the research and development community is continually improving and developing new systems which affect every weapon. A number of specialized sources in this field are worth mentioning. The basic sources on communications and electronics are again the Jane's books. *Jane's Military Communications* (New York: Franklin Watts) and *Jane's Weapons Systems* are authoritative sources of information on programs and developments. A number of the DMS, Inc. products (see Table 25) specifically *"AN" Equipment, Electronic Systems, Military Laser & EO, Electronic Warfare, Radar and Sonar, Military Communications (Europe), Code Name Handbook, Command/ Control/ Communications, Computers, Fire Control Systems,* and *Navigation/Guidance* are the basic sources on the products and development in these fields. The Frost & Sullivan *Market Reports* are also basic resources reporting developments, competition, opportunities and producers in the various communications and electronics disciplines.

There is an active military journal sub-field dealing with communications and electronics. The *Army Communicator, Communicator, Defense Electronics, Electronic Warfare Digest, Military Electronics, Journal of Electronic Defense, Counter-*

measures, and *Signal* are the main periodicals. Other military weapons periodicals also report developments in this field.

The sources on electronic warfare (EW) include two excellent reference works: *Principles of Electronic Warfare* (Robert J. Schlesinger, Los Altos, CA: Peninsula Publishing, 1979) and *The International Countermeasures Handbook* (Harry F. Eustace, ed., New York: Franklin Watts, 1975-). *Principles of Electronic Warfare* presents an overview of electronic warfare technology and tactics, including electronic intelligence. *The International Countermeasures Handbook* contains a review of EW funding, procurement and RDT&E, U.S. and foreign EW companies and programs, nomenclature and characteristics, international lexicon and code names, as well as a review of Soviet and Chinese EW systems and programs. It contains a number of articles on EW operations and systems and a bibliography.

B. REGIONAL AND COUNTRY DEFENSE ISSUES

General sources on military forces and country profiles (Sections IIB and VA) have already been discussed in this guide as an introduction to the sources which are transnational in nature and offer comparative data and information. In this section, sources on defense issues in regions of the world and individual countries are presented (European and Soviet regional and country issues are discussed in the next two sections). Many of the sources presented specifically cover military and strategic affairs. In each area, periodicals and journals, bibliographic, indexing and reference and descriptive sources are discussed.

When developing a strategic profile of a country or region, the basic reference sources in Sections IIB and VA are essential for initial data and reference. Sources like Army *Area Handbooks/Country Studies* and *Background Notes on the Countries of the World,* as well as the reference almanacs and military comparative sources are useful for developing a country profile. Raw research also depends on the many military and strategic magazines which report developments in military forces and strategic situations, as well as the CIA translations mentioned in this section.

The impact of the military as an institution on the countries of the Third World is also an area of interest for the researcher.

An understanding of the military dimension of influence and the importance of the strategic situation is essential to strategic research. A few books have discussed the effects of the military on the developing countries. A number of books on arms sales have discussed the effects of the flow of arms. Among these, *Arms Transfers to the Third World* and *Communist Aid Activities in Non-Communist Less Developed Countries* are essential reading. *Strategic Survey* puts many events in perspective vis-a-vis the Third World. The *SIPRI Yearbooks* are also valuable resources.

Many of the works concerning the role and the effect of the military on the Third World are presented in *The Military in the Developing Countries: A General Bibliography* (Charles Kuhlman, comp., Bloomington, IN: Bureau of Public Discussion, Div. of Continuing Education, Indiana University, 1971) which contains references to 1200 titles. Works and anthologies in this area include:

The Military and Security in the Third World: Domestic and International Aspects (Sheldon W. Simon, Boulder, CO: Westview, 1978).

Military Institutions and Coercion in Developing Nations (Morris Janowitz, Chicago: Univ. of Illinois Press, 1977).

The Economics of Third World Military Expenditure (David K. Whynes, Austin, TX: Univ. of Texas Press, 1979).

Middle East Politics: The Military Dimension (J.C. Hurewitz, New York: Praeger, 1969).

Armies and Politics in Africa (Harry Bienen, New York: African Publishing Co., 1978).

The Politics of Antipolitics: The Military in Latin America (Brian Loveman and Thomas M. Davies, Jr., eds., Lincoln, NE: Univ. of Nebraska Press, 1978).

Armies and Politics in Latin America (Abraham F. Lowenthal, ed., New York: Holmes and Meier, 1976).

1. African Defense Issues: The military and strategic situation in Africa receives attention only during periods of regional confrontation and crisis affecting the superpowers, but military influence and confrontation is continually shaping the African continent. Sources on the political, economic and military situation in Africa include a number of authoritative reference works, numerous African studies journals and magazines, and specialized sources on countries and current events. Many of the raw resources for primary research are in the English language, enabling advanced research. One of the biggest problems (at least in terms of strictly military areas) is the lack of information on internal military organizations, formations, capabilities and intentions. While one can often get information on arms sales and military engagements, other types of information are rarely reported in the open literature.

(a) Reference Tools: The many various sources on Africa are traceable through a number of reference tools. Two research guides to African Affairs—*Guide to Research and Reference Works on Sub-Saharan Africa* (Peter Dungan, et al., Stanford: Hoover Institution Press, 1972) and *Sub-Saharan Africa: A Guide to Information Sources* (W.A.E. Skurnik, Detroit: Gale Research Co., 1977)—are essential works for any serious analyst as references to many other sources. The major bibliographic tool is *A Current Bibliography of African Affairs* (African Bibliographic Center, Inc. (Wash.), Farmingdale, New York: Baywood Publishing Co.), a quarterly journal containing bibliographic review articles, book reviews, and subject and author classifications of articles on general African, country and regional topics. Another excellent tool is *African Abstracts* (London: International African Institute, 1950-), a quarterly index and abstracts of scholarly articles on Africa. *Africa Today* contains notes on recent publications listing many new monographs and books on U.S. policy vis-a-vis Africa and African foreign policies; and *African Affairs* and the *Africana Journal* provide good source material with review articles, bibliographies and excellent source notes. Other bibliographic sources are *Africa: Problems and Prospects* (A Bibliographic Survey) (DA Pam 550-17-1) (DA, Army Library, Wash., D.C.: GPO, Dec. 1977); *Africa: Selected References,* (Maxwell AFB, AL: Air University, January 1961-) (Supplement issued annually in January); and *Foreign Affairs Research Special Papers Available: Africa, Sub-Sahara* (Dept. of State, Wash., D.C.: GPO).

A number of excellent yearbooks on African affairs are valuable reference tools. *Africa Contemporary Record: Annual Survey and Documents* (Colin Legum and John Drysdale, London: Africa Research, Ltd./New York: Africana, 1968/69-) is the best military and strategic reference annual on Africa. It contains a review of the international relations of the African countries with references to major military documents. Country-by-country surveys and essays on current issues in foreign relations are also included. All of the material is footnoted, a valuable but rare phenomenon among annuals. *Africa South of the Sahara, 19__- __*(London: Europa Publications, 1971-), *Africa, 19__* (New York: Africana Publishing Corp., 1969-), and

Africa Yearbook and Who's Who, 19__(London: Africa Journal, Ltd., 1976-) also provide background information on African nations and regional organizations, with Who's Whos and statistical data. *Africa South of the Sahara* also contains a listing of research institutes concerned with Africa, and a bibliography of periodicals. The *Africa Yearbook* contains a chronology.

(b) Current Issues and Events: Sources on current issues and events in Africa which go beyond what is presented in the mass media are varied and extensive. Table 31 lists the various journals and magazines dealing with Africa. Magazines like *African Recorder* and *Africa Diary* are news services which summarize African events and contain excerpts from the world press. *Africa* and *New African* are general news magazines dealing with political, economic and international developments. *Strategy Week* is a Copley newsletter reporting strategic developments and arms sales relating to Africa.

Advanced and primary information on current events can be obtained from the two CIA services of the Foreign Broadcast Information Service (FBIS) and the Joint Publications Research Service (JPRS). *FBIS Daily Report, Vol. VIII: Sub-Saharan Africa* (FBIS, Springfield, VA: NTIS) contains translations of African broadcasts and editorials, and is available on a daily subscription basis from the NTIS. *Translations of Sub-Saharan Africa* (JPRS, Springfield, VA: NTIS) contains translations and

---

TABLE 31
AFRICAN JOURNALS AND MAGAZINES

Africa	Horn of Africa
Africa Currents	Issue
Africa Diary	Journal of Modern
Africa News	African Studies
Africa Report	New African
Africa Today	Paratus
African Abstracts	South Africa International
African Affairs	South African Journal
African Recorder	of African Affairs
Africana Journal	Southern Africa
Armed Forces	Strategy Week
Assegai	To the Point
Defence Africa	Washington Notes on Africa

excerpts from African language journals and specialized publications dealing with economic, political, military and international developments in Africa. It is also available through subscription.

A number of sources on U.S. policy and relationships with Africa put current events and issues into perspective. The annual foreign assistance hearings outline overall U.S. security and assistance relations with Africa. The Summer/Fall 1978 issue of *Issue* included a special report entitled "A Survey of the U.S. Government's Investments in Africa" which compiled information on U.S. presence, investments and aid. The best overall book on U.S. involvement in South Africa is *U.S. Military Involvement in Southern Africa* (Western Massachusetts Association of Concerned African Scholars, Boston: South End Press, 1978). Two conservative views of U.S. relations with Africa are presented in "U.S. Security Interests and Africa South of the Sahara" (Bruce Palmer, Jr. in *AEI Defense Review* (now *AEI Foreign Policy and Defense Review,*) Vol. II, No. 8, 1978), and "Options for U.S. Policy in Africa" (Helen Kitchen, ed., in *AEI Foreign Policy and Defense Review,* Vol. I, No. 1, 1979). The Africa Fund (198 Broadway, New York, NY 10038) is the source of many publications on Africa. The "Southern Africa Literature List," lists pamphlets and reprints available from the Africa Fund. The CRS has prepared two Issue Briefs entitled *Africa Policy* (Wash., D.C.: CRS (IB 78056), 1978) and *Africa: Soviet/Cuban Role* (Wash., D.C.: CRS (IB 78077), 1978), which contain good background information. Finally, there is a bibliography of U.S. government writings about Africa—*The U.S. and Africa: Guide to Official Documents and Government-Sponsored Publications on Africa, 1785-1975* (Julian W. Witherell, Library of Congress, Wash., D.C.: GPO, 1978)—which is the only bibliography of its kind and provides an exhaustive index of official and government sponsored documents and studies relating to U.S. policy towards Africa.

Other sources include *Ethiopia Embattled: A Chronology of Events in the Horn of Africa, 1 July 77-30 Mar. 78* (Wash., D.C.: DIA (DDB-2200-44-78), 1978) and *Ethiopia-Somalia* (Wash., D.C.: CRS (IB 78019), 1978), which describe events and problems in the Horn of Africa. Valuable sources on South Africa providing raw and background information include an irregular White Paper: *South Africa: DOD White Paper on Defense and Armament Production* (Cape Town: South African DOD); and the two South African military magazines: *Armed Forces* (South African military periodical) and *Paratus* (Official Magazine of the South African Defence Force).

2. Asian Defense Issues: Much more material is available on strategic and military affairs in Asia and the Pacific than on any other region of the Third World. A number of regional military conflicts or potential conflicts, the presence of major regional military powers, and strong U.S. and Soviet interest in the region account for the volume of writing and the interest in the area. Much of the writing on Asian security issues, specifically on the Indian-Pakistan situation, Korean peninsula issues, and Australian defense, appears in English. Traditional U.S. and British scholarly interests in this region also provides extensive resources.

(a) Reference Tools: A valuable contemporary research guide on Asian affairs is the *Scholar's Guide to Washington, D.C. for East Asian Studies* (Hong N. Kim, Wash., D.C.: Smithsonian Institution Press (Wilson Center), 1979), which lists organizations, collections and publications in the area of Chinese, Japanese and Korean studies. *America in Asia, Volume I, Research Gide on U.S. Economic Activity in Pacific Asia* (Hong Kong: Asia/North America Communications Center, 1979) is a compilation of sources of information on U.S. relations, influence, and investment in Asian countries. Other valuable sources for tracing material written on this area include a number of bibliographies:

"Bibliography of Asian Studies" appears annually in the October issue of the *Journal of Asian Studies.*
Asia: Southeastern: Selected Unclassified References (Maxwell AFB, AL: Air University Lib., Dec. 1978), a collection of sources on all of the countries of Southeast Asia including foreign policy and military references.
The Indian Ocean in International Politics: Soviet-U.S. Rivalry and the Balance of Power: A Selected Bibliography (Felix Chin, Chicago, IL: Council of Planning Librarians (Exchange Bibl. #1501), 1978?), an annotated bibliography with chronology.
South Asia and the Strategic Indian Ocean: A Bibliographic Survey of Literature (Dept. of the Army, Army Library, Wash., D.C.: GPO, 1973), an annotated selective bibliography.
Foreign Affairs Research Special Papers Available: East Asia and Pacific Area (Dept. of State, Wash., D.C.: GPO).

Annuals on Asian and Pacific affairs offer reviews of basic statistics and developments, as well as other quick reference items. Three yearbooks—*The Pacific Islands Yearbook* (Sydney,

Australia: Pacific Publications, 1932-), *Southeast Asian Affairs* (Paris, France: Institute of Southeast Asian Affairs, 1973-), and *Far East and Australasia (A Survey and Directory of the Pacific)* (London: Europa Publications, 1968-)—offer country surveys, with the *Pacific Islands Yearbook* and *Far East and Australasia* also containing a Who's Who. *Far East and Australasia* contains a review of regional and international organizations, a directory of institutes and associations studying the Orient and a bibliography of periodicals on Asia and the Pacific. *The Asia Yearbook* (Hong Kong: Far East Economic Review, 1959-), another annual, is a good source for a country-by-country review of events, including foreign relations. The *Pacific Defense Reporter Yearbook* (Dennis Warner, ed., Church Point, Australia: PY Logistics & Holdings, Ltd., 1978-) is a specialized annual which includes an assessment of Asian, Middle-Eastern, African and South Pacific defense as well as descriptions of developments in those areas. It is valuable for its discussion of Australian and New Zealand military and strategic affairs. *Asian Security 19__*(Tokyo, Japan: Japanese Research Institute for Peace and Security, 1979-) is a new annual with information on Asian military issues, Japanese defense and the military balance, and includes a chronology.

Two other sources that offer good military analysis and background information on Asian military affairs are *Guide to Far Eastern Navies* and *The Military Equation in Northeast Asia* (Stuart E. Johnson and Joseph E. Yager, Wash., D.C.: Brookings Institution, 1979). Both contain data and analysis on Chinese, Taiwanese, Japanese, Korean and U.S. and Soviet military forces and the regional military balance.

(b) Current Issues and Events: Table 32 is a listing of the various journals and magazines that cover Asian and Pacific defense issues on a regular basis. General news of Asia is provided in *Asian Recorder* (weekly digest of events in Asia), *Asiaweek, Far Eastern Economic Review* (the *Time* magazine of Asia), and *Japan Times Weekly. Strategy Week* is a Copley newsletter which covers military and foreign policy developments including arms sales. Three magazines—*Asian Defence Journal, Pacific Defense Reporter* and *Vikrant*—specialize in Asian military affairs. *Asia Monitor* is an excellent source for internal developments in Asian countries and U.S. economic, political and military relations. It is well-researched and documented.

Primary information is available through the CIA-produced

TABLE 32
ASIAN JOURNALS AND MAGAZINES

AMPO
Asia Monitor
Asia Pacific Community
Asia Quarterly
Asia Research Bulletin
Asian Affairs
Asian Defence Journal
Asian Perspective
Asian Recorder
Asian Survey
Asiaweek
Australian Foreign
   Affairs Record
Australian Journal of
   Politics and History
The Australian Journal
   of Defence Studies
Australian Outlook
Bulletin of Concerned
   Asian Scholars
China Quarterly
China Letter
China Report
Contemporary China
Defence Force Journal
Defence Journal
Defence Management
Defence Science Journal

Far East Economic Review
IDSA Journal
India Quarterly
Japan Quarterly
Japan Times Weekly
Journal of Contemporary Asia
Journal of Asian Studies
Journal of Southeast Asian Studies
Korea and World Affairs
Korea Observer
Military Digest (India)
Military Digest (Pakistan)
Modern China
National Security Review
Pacific Affairs
Pacific Defence Reporter
Pacific Research
Pakistan Army Journal
Pakistan Horizon
Pioneer: Singapore Armed
   Forces News
Southeast Asia Chronicle
Strategic Studies
Strategy Week
USI Journal
Vantage Point
Vikrant: Asia's Defense
   Journal

translations of the FBIS and the JPRS. The JPRS has four subscription translation services dealing with Asia:

   Translations on Mongolia (JPRS, Springfield, VA: NTIS)
   Translations on Vietnam (JPRS, Springfield, VA: NTIS)
   Translations on South and East Asia (JPRS, Springfield, VA: NTIS)
   Problems of the Far East (JPRS, Springfield, VA: NTIS)

FBIS also has a daily translation service of broadcasts which is available from the National Technical Information Service: *FBIS Daily Report: Vol. IV: Asia/Pacific* (FBIS, Springfield, VA: NTIS).

U.S. relations with the Asian nations are discussed annually in the foreign assistance hearings and in the Defense Department hearings. Since the United States maintains large forces in the region and has an active military assistance program, there is usually a wealth of information available on the region.

(c) Australian Defense Issues: A number of studies and public documents are available on Australian defense and military affairs. Annually, the Australian Defense White Papers—the *Defence Report* (Canberra: Australian Dept. of Defense), *White Paper on Australian Defence presented to Parliament by the Minister for Defence* (Canberra: GPS), and *Australian National Accounts* (Canberra: Bureau of Statistics)—are presented by the Australian government. These include a review of Australian military programs and funding and an Australian strategic outlook. A foreign affairs review, the *Australian Foreign Affairs Record* is also published monthly by the Department of Foreign Affairs.

Background information on Australian military and strategic affairs is available from a number of sources:

The Defence Force of Australia (Army Quarterly and Defence Journal, Tavistock, Devon, U.K., 1977), basic information and description of the Australian military.

Australian Defence Resources: A Compendium of Data (Jolika Tie, et al., Canberra: ANU Press, 1978), supplementary data and information not contained in the Defence White Papers.

Australia and the Indian Ocean Region (Report from the Senate Standing Committee on Foreign Affairs and Defence, Canberra: GPS (Catalog No. 76 3930 8), 1977).

The Defence of Australia: Fundamental New Aspects: The Proceedings of a Conference Organized by the Strategic and Defence Studies Centre, The Australian National University, October 1976 (Robert O'Neill, ed., Canberra: ANU Press, 1977).

Australian Aviation Year Book (John Stackhouse, Sydney: Pacific Yearbooks, 1974-), history and current status of the Australian DOD, Royal Australian Air Force, Army and Navy air arms, including current aircraft and Who's Who.

Australian Defence Equipment Catalogue (3d Ed.) (Melbourne: Peter Isaacson Publications, 1979), Australian military equipment available for sale.

A Suitable Piece of Real Estate: American Installations in

Australia (Desmond Ball, Australia: Hale and Iremonger, 1980).

The Strategic and Defence Studies Centre of the Australian National University conducts research on the issue of Australian defense and publishes studies and reports on the subject as well as periodic *SDSC Working Papers*. The Centre should be contacted when doing research specifically related to Australian defense.

(d) Indian Defense Issues: The Indian material on military affairs and Indian-Pakistani military relations is also in English. Two research tools useful for tracing sources on Indian foreign and military policy are "India and World Affairs: An Annual Bibliography," appearing in the January-March issue of *International Studies* and *India Quarterly* magazine. Both of these resources have extensive references to works on all aspects of Indian military and foreign policy. *India Quarterly* contains extensive reviews of Indian books and material on foreign and military policy.

A number of annuals dealing with military affairs are published in India. *The Military Year Book* (S.P. Baranwal, ed., New Delhi, India: Guide Publications, 1965-) and the *Chanakya Defence Annual* (Allahabad, Chanakya Publishing House) are unique in-depth resources on the Indian military. *The Indian Armed Forces Year Book* is an excellent source, containing articles on the Indian military and the regional and global military situation.

Government publications include the *Annual Report* (New Delhi: Ministry of Defence, 1958-) of the military; *Foreign Affairs Record,* the monthly review of Indian foreign policy including documents, releases and speeches; and other Indian military periodicals:

Defence Management
Defence Science Journal
IDSA Journal (India)
Military Digest (India)
United Services Institute Journal

A number other other periodicals also published in India (see Appendix A)—specifically, international relations and strategic journals—are also of interest for their coverage of the Indian subcontinent. One outstanding source is the March-April 1979 issue of *China Report,* which contains a comprehensive review of Indian-Chinese relations from 1947-1979 with a chronology and bibliography.

(e) Japanese Defense Issues: A number of excellent sources are available in English on Japanese military and strategic affairs. The best research tool is *Japan Quarterly,* which contains a quarterly chronicle of Japanese politics and international relations, including an excellent chronology, book reviews, and announcements of recent publications on Japan. An annual reference work in the English language is the *Japan Annual of International Affairs* (Tokyo, Japan: Japanese Institute of International Affairs, 1964-) which covers developments in Japanese and Asian foreign policy.

The Japanese government also makes a number of official documents available in English:

White Paper of Japan (*Annual Abstracts of Official Reports and Statistics of the Japanese Government*) (Tokyo, Japan: Japanese Institute of International Affairs), abstracts of official reports and statistics of the Japanese government including translations of the official White Papers of the Defence Agency and the Foreign Affairs Ministry.

The Defense of Japan (Tokyo, Japan: Japanese Defense Agency), a defense White Paper.

Japan 3-Defense Forces Equipment Yearbook (Tokyo, Japan: Defense Agency).

Information Bulletin (Tokyo, Japan: Public Information Bureau, Ministry of Foreign Affairs), compilation of releases of the Japanese government for the year.

Current events relating to Japan are discussed in *Japan Times Weekly* as well as the other general Asian news magazines. *Translations on Japan* (JPRS, Springfield, VA: NTIS) is another of the CIA's services available on subscription which translates articles of importance from Japanese.

Some recent works on Japanese military and strategic affairs are:

The Postwar Rearmament of Japanese Maritime Forces, 1945-1971 (J.E. Auer, New York: Praeger, 1973).

United States-Japan Security Relationship, The Key to Asian Security and Stability (A Report of the Pacific Study Group to the Senate Armed Services Committee, 96-1, Wash., D.C.: GPO, March 1979).

The Military Equation in Northeast Asia.
Guide to Far Eastern Navies.

Japan's Contribution to Military Stability in Northeast Asia (prepared by ACDA for the Subcommittee on East Asian and Pacific Affairs of the Senate Foreign Relations Committee, 96-2, Wash., D.C.: GPO, June 1980).

(f) Korean Defense Issues: Most of the writing and interest in the Korean peninsula in military and strategic affairs revolves around the U.S. presence and commitment in the region and the military balance between the two countries. President Carter's abortive decision to withdraw troops from South Korea also resulted in a rash of writings. Background information on the present situation in Korea is available from the posture statements and hearings where U.S. forces and programs relating to Korea are normally discussed, and in a number of background sources:

Deterrence and Defense in Korea: The Role of U.S. Forces (Ralph N. Clough, Wash., D.C.: Brookings, 1976), an examination of the overall role of U.S. forces, and the military and political significance of their presence.
U.S. Policy Toward Korea: Analysis, Alternatives and Recommendations (Nathan N. White, Boulder, CO: Westview Press, 1979), a good discussion of U.S.-Korean relations and policy options.
The Threat to South Korea (Wash., D.C.: DIA, January 1977).
Guide to Far Eastern Navies.
The Military Equation in Northeast Asia.

Current events relating to Korea are discussed in the general Asian news magazines. *Translations on North Korea* (JPRS, Springfield, VA: NTIS), a CIA translation service, is an excellent source of raw information on this area. An excellent journal, *Korea and World Affairs* (A Quarterly Review) contains articles about Korean peninsula military and strategic affairs, and also has an excellent chronology of North and South Korean developments.

Finally, there are a number of sources relating to the Carter decision to withdraw troops from Korea. These sources include varying amounts of regional analysis, military balance statements and policy discussion of the U.S. commitment and the role of the U.S. presence. *Korea: U.S. Troop Withdrawal* (Wash., D.C.: CRS (IB 79053), 1979) and *Force Planning and Budgetary Implications of U.S. Withdrawal from Korea* (CBO, Wash., D.C.: GPO, April 1978) are the best sources for background data. The following Congressional hearings and documents also examined the issue and should be consulted for research:

Hearings on Review of the Policy Decision to Withdraw United States Ground Forces from Korea (Hearings before the Investigations Subcommittee of the House Armed Services Committee, 95-1/2, Wash., D.C.: GPO, 1978).
Review of the Policy Decision to Withdraw U.S. Ground Forces from Korea (Report of the Subcommittee on Investigation, House Armed Services Committee, 95-2, Wash., D.C.: 1978).
U.S. Troop Withdrawal from the Republic of Korea (Hubert H. Humphrey and John Glenn, A Report to the Senate Committee on Foreign Relations, 95-2, Wash., D.C.: GPO, 1978).
U.S. Troop Withdrawal from the Republic of Korea (John Glenn, Report to the Senate Committee on Foreign Relations, 96-1, Wash., D.C.: 1979).
Intelligence Reassessment of Impact of Withdrawal of U.S. Troops in Korea (Hearings before the House Armed Services Committee, 96-1, Wash., D.C.: GPO, 1979).
Report on Impact of Intelligence Reassessment of Withdrawal of U.S. Troops in Korea (Report of the House Armed Services Committee, 96-1, Wash., D.C.: GPO, 1979).

(g) Other Country Defense Issues: Information on defense issues affecting countries other than the ones specifically mentioned above is available to a lesser extent. Some has been written on Malaysia and Singapore, most recently *The Defence of Malaysia and Singapore* (David Hawkins, London: RUSI (Defence Studies), 1978), an excellent monograph. *Foreign Affairs Malaysia* (Ministry of Foreign Affairs, Malaysia, Ibrahim Bin Johari, P15, Director General, Peninsula, Kuala Lumpur, Malaysia: Govt. Printing Dept.) is another source, providing a quarterly review of Malaysian foreign policy including a diary of events.

Some information on Taiwanese defense issues is included in *The Military Equation in Northeast Asia,* the *Guide to Far Eastern Navies,* and a CRS report, *Taiwan's Future: Implications for the United States* (Wash., D.C.: CRS (IB 79101), 1979).

Most documents and publications on Pakistani military affairs are also in English. *Pakistan Horizon* is an excellent tool for finding publications and documents with an extensive presentation on Pakistani military and foreign policy. It also contains an excellent chronology. A number of English language military publications are of interest to the advanced researcher: *Defence Journal, Military Digest, Pakistan Army Journal,* and *Strategic Studies.*

Finally, there are some New Zealand sources worth mention.

The annual White Paper of the New Zealand military, *New Zealand, Ministry of Defense, Review of Defense Policy* ("Defense Review") (Minister of Defense, Wellington, NZ) is an essential source. The *New Zealand Foreign Affairs Review,* the quarterly publication of the Ministry of Foreign Affairs, reviews New Zealand foreign policy, and includes statements, releases, articles and a country-by-country review of policy statements and events. *Pacific Defence Reporter* reports developments in the New Zealand military.

3. Chinese Defense Issues: Recently more information has become available on Chinese military and strategic affairs, but writing and analysis is still sparse. A good, though somewhat limited and selective bibliography dealing with the military field is *China: An Analytical Survey of Literature* (DA Pam 550-9-1) (DA, Army Library, Wash., D.C.: GPO, 1978). It contains some notes on further research and references that are useful.

(a) Reference Tools: Reference tools on China include *Contemporary China,* with extensive and excellent source notes and bibliography; "CIA Publications on China," (Andrew J. Nathan, *Contemporary China,* Spring 1979), a short review essay; and *Foreign Affairs Research Special Papers Available: People's Republic of China* (Dept. of State, Wash., D.C.: GPO), a bibliography of external research reports of the Department of State. Most of the major international relations and military bibliographies mentioned in earlier sections also contain sections on China.

Two excellent reference works on China are *China Facts and Figures Annual* (John L. Scherer, ed., Gulf Breeze, FL: Academic International Press, 1978-) and *PRC: A Handbook* (Harold C. Hinton, ed., Boulder, CO: Westview, 1979). *China Facts and Figures Annual* is a new annual compendium of data on government, the military, economics, demographics, institutions and personalities. It is sourced and relies on CIA and other government documents as well as a wide range of private materials and contains a chronology. The coverage from issue to issue varies, so each issue is complementary. *PRC: A Handbook* contains background articles on various aspects of China, including political system, economic development, science and technology, the military, and foreign relations.

(b) Current Issues and Events: The primary sources on Chinese current events are the CIA translations. *FBIS Daily Report: Vol. I: PRC* (FBIS, Springfield, VA: NTIS) translates

broadcasts and the various JPRS publications below translate journals and documents:

Translations on PRC: Sociological, Economic, Military, Scientific and Technical Information (JPRS, Springfield, VA: NTIS).
Translations on PRC: Scientific Abstracts (JPRS, Springfield, VA: NTIS).
Translations on PRC: Agriculture (JPRS, Springfield, VA: NTIS).
Translations on PRC: Plant and Installation Data (JPRS, Springfield, VA: NTIS).
Translations from "Red Flag" (JPRS, Springfield, VA: NTIS).

In addition, a number of journals dealing with China follow developments in Chinese military and strategic affairs:

China Quarterly
China Letter
China Report
Contemporary China
Modern China

(c) Chinese Foreign Policy: General works on Chinese foreign policy include:

China and Asia: An Analysis of China's Recent Policy Towards Neighboring States (Report prepared for the Subcommittee on Asian and Pacific Affairs, House Foreign Affairs Committee, by the CRS, LC, 96-1, Wash., D.C.: GPO, March 1979).
The United States and China (4th Ed.) (John King Fairbank, Cambridge, MA: Harvard University Press, 1979).
China-U.S. Relations (Wash., D.C.: CRS (IB 76058), 1976).
China-U.S.-Soviet Relations: Should We Play the China Card? (Wash., D.C: CRS, (IB 79115), 1979).

The Joint Economic Committee hearings, *Allocations of Resources in the Soviet Union and China* are also excellent sources on Chinese developments in the foreign policy and military fields.

(d) The Chinese Military: Some of the analyses of the Chinese military as an institution and of Chinese military policy are worth reference. The DIA *Handbook on the Chinese Armed Forces* (DIA, (DDI-2680-32-76), Wash., D.C.: GPO, 1976) is the basic reference work on the subject, full of background information. It has also been reproduced as *The Chinese Armed Forces Today* by Prentice Hall and is sold commercially. "China's Air and Missile Forces," Special Issue, *Flight International* (22 Sept. 79), is a more in-depth and critical update of certain parts of the DIA study. Other sources like *Guide to the Far Eastern Navies* and *The Military Equation in Northeast Asia* also provide valuable information. Annually, the *Joint Chiefs of Staff Posture Statement* provides an assessment and analysis of the Chinese military, particularly the strategic capabilities. The Chinese military has received much attention recently in the military magazines, particularly relating to the status of the military and its modernization. The general indexes should be consulted for these articles.

4. Latin American Defense Issues: Very little material is available on Latin American military and strategic affairs in the English language. In fact, very little attention in the military analysis community is given to the military situation in Latin America. What interest there is deals with Cuba and the Caribbean, where the United States has a military presence and interests. But on the South American countries, where militaries continue to grow with their own production capabilities and where there are a number of potential conflicts, very little is written.

(a) Reference Tools: Some bibliographic sources of writings on Latin America are *Latin America and the Caribbean: An Analytical Survey of Literature* (DA Pam 550-7-1) (DA, Army Library, Wash., D.C.: GPO, 1975), a selective annotated bibliography; *Handbook of Latin American Studies* (Gainesville, FL: Univ. of Florida Press, 1935-), an annual annotated bibliography in two volumes (one Social Sciences, one Humanities) including essays and bibliographic articles; and *Foreign Affairs Research Special Papers Available: American Republics* (Dept. of State, Wash., D.C.: GPO). *The Handbook of Latin American Studies* is the best English language source of writings on Latin American militaries.

Two research guides are also available on Latin American studies: *Scholar's Guide to Washington, D.C.: Latin American and Caribbean Studies* (Michael Grow, Wash., D.C.: Smithsonian Institution Press (Wilson Center), 1979) and *Latin Americana Research in the United States and Canada: A Guide and Directory* (Robert P. Hard, Chicago: American Library Assn., 1971). *The Scholar's Guide* lists collections, organizations and publications in the Washington area, and includes a bibliography and index.

Latin Americana Research is a directory of special collections and libraries and other specialized sources.

Two annuals on Latin America are valuable reference tools. *The Caribbean Yearbook of International Relations* (Leslie F. Manigat, ed., Winchester, MA: Sijhoff, 1975-), contains articles on political, geopolitical and international developments in the countries of the Caribbean, with a chronology and index. *Latin America Annual Review & The Caribbean, 19__* (London: World of Information, 1979-) contains regional reviews and country surveys, a review of the past year and an analysis of economic trends.

Finally, the various Latin American journals are also good reference tools. *Latin American Research Review,* for instance, is an excellent research tool with extensive book reviews, research reports, notes and bibliographic articles on current issues.

(b) Current Issues and Events: The primary sources on current Latin American developments are the CIA translations from FBIS and JPRS. *FBIS Daily Report: Vol. VI: Latin America* (FBIS, Springfield, VA: NTIS) and *Translations on Latin America* (JPRS, Springfield, VA: NTIS) are excellent sources which cover the major issues and developments. The various periodicals which deal with Latin America are:

> *Caribbean Review*
> *Cuban Studies*
> *Defense Latin America*
> *Inter-American Economic Review*
> *Latin American Political Report*
> *Latin American Digest*
> *Latin American Research Review*
> *NACLA Report on the Americas*
> *Strategic Latin American Affairs*

Strategic Latin American Affairs is a Copley newsletter which reports military and strategic developments and arms sales.

U.S. relations with Latin America are discussed annually as part of the foreign assistance and military assistance hearings. A bibliography of U.S. relations with Latin America is *U.S. Intervention in Latin America: A Selected and Annotated Bibliography* (Miles D. Wolpin, New York: American Institute for Marxist Studies, 1971).

Two recent works on the military in Latin America are *The Politics of AntiPolitics: The Military in Latin America* and

Armies and Politics in Latin America. Both of these books examine various political, economic and historical factors affecting military roles in Latin America and military influence.

5. Middle East Defense Issues: The Arab-Israeli conflict and the issue of Persian Gulf security and natural resources have made the Middle East an important strategic region, stimulating much written material on the military and strategic affairs of the area. The information available on the military forces of the region is not very good, and the sources discussed in Sections IIB and VA are probably the best sources for force structure and organizational data.

(a) Reference Tools: Reference tools dealing with the Middle East include a number of excellent (and new) sources, indicating wide interest in the region. Two excellent research guides detail the many sources on the region. *The Modern Middle East: A Guide to Research Tools in the Social Sciences* (Reeva S. Simon, Boulder, CO: Westview, 1978) is the most up-to-date and comprehensive, with a good subject index and notes on references, sources, abstracts, periodicals and bibliographies. *International and Regional Politics in the Middle East and North Africa: A Guide to Information Sources* (Ann Schulz, Detroit, MI: Gale Research, 1977) is also good, with a restrospective bibliography and reference sources on many aspects of Middle Eastern international relations.

Three bibliographies of Middle East sources are *The Middle East: Abstracts and Index* (Pittsburgh, PA: Northumberland Press, 1978-), *The Middle East Journal,* and *Foreign Affairs Research Special Papers Available: Near East, South Asia and North Africa* (Dept. of State, Wash., D.C.: GPO). *Middle East Abstracts and Index* is a new quarterly index of periodical articles covering 16 nations and the "Arab-Israeli Conflict," "General Interest," and the "Arab World." The titles are well-indexed and come from a wide variety of sources, including 355 periodicals, books, government documents and dissertations. The *Middle East Journal* contains a comprehensive listing of new books and a bibliography of periodical literature in general Middle East political and international affairs. Each issue also includes an excellent Arab-Israeli and country-by-country chronology.

Three annual reference works are valuable for quick reference and analysis. *The Middle East and North Africa, 19__ - __* (London: Europa Publications, 1948-) contains regional and country surveys with basic statistical data. It also contains a reference section with a listing of research institutes interested in Middle

Eastern affairs, a bibliography of periodicals, and articles on the arms trade with the Middle East. *Middle East Annual Review* (Michael Field, ed., Essex, U.K.: Middle East Review Co., Ltd., 1978) contains background articles by journalists on economic and political issues of Middle East nations and the region. *Middle East Contemporary Survey* (Colin Legum and Haim Shaked, eds., London: Holmes & Meier Publishers, Inc., 1978) is the most useful for military and strategic reference with articles on political, military, international, economic and social issues including strategic surveys and country-by-country reviews.

(b) Current Issues and Events: The CIA translation services are the best primary sources of information on current developments in the Middle East. *Translations on Near East and North Africa* (JPRS, Springfield, VA: NTIS) and *FBIS Daily Report: Vol. V: Middle East & North Africa* (FBIS, Springfield, VA: NTIS) both report military and strategic developments and other events which never make the mass media. *MEED Arab Report* (formerly *Arab Report and Record*), *Middle East Economic Digest* and *Middle East Record* are all summaries of events and developments in the Middle East. The periodicals which deal with the Middle East region are:

> *Arabia and the Gulf*
> *Born in Battle*
> *International Problems*
> *Jerusalem Journal of International Relations*
> *Journal of Palestine Studies*
> *MEED Arab Report*
> *MERIP Reports*
> *Middle East*
> *Middle East Economic Digest*
> *Middle East International*
> *The Middle East Journal*
> *Near East Report*
> *Strategy Week*

Strategy Week is a Copley newsletter which reports military and strategic affairs, including arms sales.

U.S. relations with the region are reviewed annually in the hearings on foreign and military assistance. Various writings since 1979 dealing with U.S. rapid deployment forces and the Carter Doctrine also outline U.S. policy vis-a-vis the Middle East.

(c) The Arab-Israeli Conflict: The Arab-Israeli conflict is

the most prominent military and strategic issue in the region. The conflicts between the Arabs and the Israelis, the threat of confrontation in the region, arms control and the peace process are topics of continual analysis. Two bibliographies covering much of what has been written are *Palestine and the Arab-Israeli Conflict: An Annotated Bibliography* (Beirut, Lebanon: Institute for Palestine Studies, 1974) and *The Arab-Israeli Conflict: A Historical, Political, Social and Military Bibliography* (Ronald M. Devore, Santa Barbara, CA: ABC-Clio Press, 1976). *Middle East Abstracts and Index* and the *Middle East Journal* also contain sections on the Arab-Israeli conflict, and are valuable contemporary resources. Another excellent current reference tool is the *Journal of Palestine Studies*. Each issue contains five bibliographic review sections covering the Middle East and the Arab-Israeli conflict: "From the Arab Press," "Arab Reports and Analysis," "Views from Abroad," "Documents and Source Material," and "Periodicals in Review."

(d) Persian Gulf Security: Security issues in the Persian Gulf region are increasingly becoming a cause celebre, particularly as a result of the revolution in Iran and the Soviet invasion of Afghanistan. A number of works have bene done on the region and it is assumed many more will be done, including numerous Congressional hearings and studies. Some sources which give a good amount of background information are:

Military Forces in the Persian Gulf (Alvin J. Cottrell and Frank Bray, CSIS (Washington Papers Vol. VI, No. 60), Berkeley, CA: SAGE, 1978).

"American Policy Options in Iran and the Persian Gulf" (Robert J. Pranger and Dale R. Tahtinen, in *AEI Foreign Policy and Defense Review*, Vol. I, No. 2, 1979).

Petroleum Imports from the Persian Gulf: Use of U.S. Armed Forces (Wash., D.C.: CRS (IB 79046), 1979).

The Clouded Lens: Persian Gulf Security and U.S. Policy (James H. Noyes, Stanford, CA; Hoover Institution Press, 1979).

U.S. Security Interests and Policies in Southwest Asia (Hearings before the Subcommittee on Near East and South Asia of the Senate Foreign Relations Committee, 96-2, Wash., D.C.: GPO, 1980).

U.S. Security Interests in the Persian Gulf (Report of a Staff Study Mission to the Persian Gulf, Middle East and North Africa, to the House Foreign Affairs Committee, 97-1, Wash., D.C.: GPO, 16 March 1981).

U.S. Interest in and Policies Toward the Persian Gulf, 1980

(Hearings before the Subcommittee on Europe and Middle East, House Foreign Affairs Committee, 96-2, Wash., D.C.: GPO, 1980).

C. THE SOVIET UNION, EASTERN EUROPE AND THE WARSAW PACT

Hard information on the Soviet military is scarcely available other than through the U.S. or Western intelligence services. Indeed, were it not for the many varied works by these organizations and the many official studies and reports drawing from this intelligence data, information of this sort would be very rare. As it is, there is ample contradiction and mis-statement to keep both sides of the issues armed with the data to back their assertions.

Research on the Soviet military and on Soviet military and strategic affairs thus becomes for some a search for the right statistics and points which support pre-conceived ideas. Even those with the capability of reading Russian find that while there is an abundance of polemical and instructive material available, it comes no where near the quality of information which appears in even one set of Congressional hearings on the U.S. military. Luckily the intelligence community and the military need to make large amounts of information available in an unclassified form in order to train its people, and to testify in support of its annual budget submission. This data and information is the heart of all of the writings that the debate uses.

1. Reference Tools: There are, first, many bibliographies to trace the volumes written on every aspect of Soviet military and strategic affairs:

"Key Works on Russia and the Soviet Union," in *USSR Facts and Figures Annual,* a continuing annotated bibliography concentrating on contemporary writing in military affairs.

U.S.S.R.: Analytical Survey of Literature (DA Pam 550-6-1) (DA, Army Library, Wash., D.C.: GPO, 1976), a selective annotated bibliography of works on Soviet military and foreign policy.

Soviet Military Capabilities: Selected References (Maxwell AFB, AL: Air University Library, Feb. 1977 (Supplement 1, July 1979)), a comprehensive bibliography on policy, doctrine, forces, capabilities and research and development.

Bibliography of Unclassified Books and Monographs on the Soviet and East European Ground Forces (DIA, (DDB-1100-164-78), Wash., D.C.: 1978).

Foreign Affairs Research Special Papers Available: U.S.S.R. (Wash., D.C.: Dept. of State, GPO).

"The Counterpart of Defense Industry Conversion in the United States: The U.S.S.R. Economy, Defense Industry, and Military Expenditures: An Introduction and Guide to Sources," (Milton Leitenburg, *Journal of Peace Research*, Vol. XVI, No. 3, 1979).

The International Relations of Eastern Europe: A Guide to Information Sources (Robin Allison Remington, Detroit, MI: Gale Research Co., 1978), a general bibliography of standard and classic sources.

Communist Eastern Europe: An Analytical Survey of Literature (DA Pam 550-8) (DA, Army Library, Wash., D.C.: GPO, 1971), a selective annotated bibliography on military and international issues.

Another reference tool is the research guide, *Scholar's Guide to Washington, D.C.: Russian/Soviet Studies* (Steven A. Grant, Wash., D.C.: Smithsonian Institution Press (Wilson Center), 1977). This guide lists collection, organizations and publications involved in Russian and Soviet studies in the Washington area.

2. Current Issues and Events: The primary sources on Soviet developments and news are the CIA translation services and a few commercial translation services. The CIA services—FBIS and JPRS—are available by subscription through the National Technical Information Service and are translations of Soviet broadcasts and writings. Table 33 lists the various JPRS and FBIS services that are done on Eastern Europe and the Soviet Union. The *Current Digest of the Soviet Press* and *Soviet World Outlook* are commercial newsletters which report articles in the Soviet press. The *Current Digest* is a weekly compilation of translations of major documents and significant articles from the Soviet press. It is accompanied by four quarterly indexes. *Soviet World Outlook* is a monthly newsletter which reports activities or announcements and articles from the Soviet press on Soviet internal affairs and Soviet-U.S. relations. *Soviet Analyst* reports developments in Soviet policy, and *Soviet Aerospace* is a newsletter which reports developments in the Soviet military.

The periodicals specializing in Soviet studies are essential

TABLE 33
CIA TRANSLATION SERVICES ON THE U.S.S.R. AND EASTERN EUROPE

Translations on U.S.S.R.:
 Military Affairs
 Economic Affairs
 Political and Sociological Affairs
 Biomedical & Behavioral Sciences
 Resources
 Industrial Affairs
 Sociological Studies
 Trade and Services
 S&T: Physical Sciences and Technology
 Space Biology and Aerospace Medicine

Translations from "Kommunist"

Translations on Eastern Europe:
 Scientific Affairs
 Political, Sociological and Military Affairs
 Economic and Industrial Affairs

FBIS Daily Report: Vol. II: Eastern Europe

FBIS Daily Report: Vol. III: Soviet Union

---

reading. Of course, all of the military magazines include articles on Soviet military activities and developments. The following magazines, however, specialize in the Soviet Union:

 Current Digest of the Soviet Press
 Problems of Communism
 Review of the Soviet Ground Forces
 The Russian Report
 Russian Review
 Slavic Review
 Soviet Aerospace
 Soviet Analyst
 Soviet Military Review
 Soviet Press: Selected Translations
 Soviet Studies

Soviet World Outlook
Studies in Comparative Communism

Writings on Soviet doctrine and strategy are another area of importance to researchers. The sources on Soviet doctrine are for the most part translated from the Russian. *Soviet Sources of Military Doctrine and Strategy* (William F. Scott, New York: Crane, Russak, 1975) discusses 168 Soviet sources on military doctrine and strategy published between 1960 and 1974. "The Soviet Military Press 1978" by John Erickson in *Strategic Review* (Summer 1979), and "Soviet Military Doctrine in the 70's" by Paul F. Walker in *Problems of Communism* (July-Aug. 1979), are two other sources.

3. *Soviet Foreign Policy and Relations:* Much has been written on Soviet-U.S. policy and Soviet policy vis-a-vis the Third World. Bound to become the classic study on the political use of Soviet military power is *Diplomacy of Power: Soviet Armed Forces as a Political Instrument* (Stephan S. Kaplan, Wash., D.C.: Brookings, 1981), which includes an appendix listing 187 incidents involving the use of Soviet armed forces from 1944 to 1979. It also includes an extensive bibliography and a list of chronologies. An excellent Congressional publication is *Soviet Policy and U.S. Response in the Third World* (Report by the CRS, LC, for the House Foreign Affairs Committee, 97-1, Wash., D.C.: GPO, March 1981). Overall Soviet policy and intentions are examined in Congressional hearings which have been held regularly to examine the state of the Soviet economy, defense spending and the allocation of resources. These hearings, first initiated by the Joint Economic Committee in 1974, analyze the size and characteristics of defense programs of the Soviet Union and the commitment of the Soviet Union to the military. They are published as continuing volumes of *Allocation of Resources in the Soviet Union and China* (Hearings before the Joint Economic Committee, Wash., D.C.: GPO, 1975-). Recently, a study of the Soviet economy, *Soviet Economy in a Time of Change* (Study prepared by the CRS, LC, for the Joint Economic Committee, Wash., D.C.: GPO, 1979), was completed which ties together the Soviet economic picture and the military picture. Characterizations of Soviet objectives and foreign policy are made throughout the hearings process on the Defense Budget and foreign assistance. The *Posture Statements* of the Secretary of Defense and the Chairman of the Joint Chiefs of Staff also characterize Soviet policy and intentions. Finally, the CIA publishes an annual study, *Communist Aid Activities in*

Non-Communist Less Developed Countries, 19__, which is a survey of economic and military aid programs of the Soviet Union with country surveys of Soviet influence and relationships.

4. Soviet and East European Military Forces: The size of the Soviet military, its deployment and operations, and its intentions are important issues in the military and strategic considerations of the Soviet Union's position in the world and its relations with the United States. Earlier, some impediments to research in this area were outlined. A good deal of information is available for most types of research, however, and the sources mentioned in this guide provide leads to a great percentage of this material. All military magazines and writings consider the Soviet military, many U.S. hearings are held on the character and intentions of the Soviet Union, and a multitude of organizations and spokesmen cover all sides of the issues. What may be difficult, especially for the novice or non-specialist researcher, is understanding how different analysts can come to differing conclusions when they are given basically the same data to work from. One way to evaluate the validity of conflicting assessments is to examine the criteria used in measuring Soviet capabilities. Do the assessments measure qualitative factors (e.g., technical sophistication, reliability, accuracy) as well as numbers? Do they cover Soviet weaknesses as well as strengths? Do they distinguish between older and newer, more- and less-capable systems in Soviet arsenals? After examining the plethora of materials on this topic, the serious researcher will develop a sensitivity to these distinctions and thus will be able to formulate his or her own assessment of the available data.

(a) Background: A number of sources present basic reference material on the Soviet and East European armed forces. The various DOD publications include the *Posture Statements* and the Congressional hearings already mentioned. These types of publications are the basis for a number of good reference works:

Soviet Armed Forces Review Annual (David R. Jones, ed., Gulf Breeze, FL: Academic International Press, 1977-), an annual compilation of articles on the branches of the Soviet Armed Forces, with charts and amplifying data including material on the Soviet defense industry, civil defense and military forces. The volumes cover military developments since 1974; the contents of each volume vary.

U.S.S.R. Facts & Figures Annual (John L. Scherer, ed., Gulf Breeze, FL: Academic International Press, 1977-), an annual compendium of much data on government, armed forces, demography, economy, and institutions. The charts and text are sourced. The coverage varies slightly from issue to issue. Contains a useful chronology.

Handbook of the Soviet Armed Forces (DIA, (DDB-2680-40-78), Wash., D.C.: GPO, Feb. 1978), an introductory survey and fact book.

The Armed Forces of the U.S.S.R. (Harriet Fast Scott and William F. Scott, Boulder, CO: Boulder, 1979), a comprehensive introduction and background text on the history and doctrine, strategy and tactics, organizations and institutions, personalities and politics of the Soviet Armed Forces.

The Military-Naval Encyclopedia of Russia and the Soviet Union (David R. Jones, Gulf Breeze, FL: Academic International Press, 1978-), a multi-volume (50 are planned) encyclopedia on all aspects of the Soviet military.

"Soviet Aerospace Almanac" Issue, *Air Force Magazine*, March, an annual assessment of the U.S.-U.S.S.R. military balance, Soviet activities, Soviet air forces, missile forces, and Soviet weapons, from the Air Force Association.

The Armies of the Warsaw Pact Nations: Organization, Concept of War, Weapons and Equipment (Friedrich Wiener, Vienna, Austria: Carl Ueberreuter Publishers, 1976), an excellent and brief reference work on armed forces of the Warsaw Pact, including weapons, strategy, doctrine and organization.

Soviet-Warsaw Pact Force Levels (John Erickson, Wash., D.C.: U.S. Strategic Institute (USSI Report 76-2), 1976), covers military organization, force structure, inventory; more useful for non-Soviet forces than Soviet forces information.

A number of reference works which concentrate on military equipment have already been discussed (see Section VA). Most of the general reference works on Armed Forces (see Sections VA and IIB) also have valuable information.

(b) Ground Forces: Background information specifically on the ground forces is also available. An introductory source is the *Handbook on Soviet Ground Forces* (FM 30-40) (DA, Wash., D.C.: GPO, 1976), which discusses various aspects of organization, history and doctrine. *The Armies of Europe Today* and *The Armies of the Warsaw Pact* also have good material. *Understanding Soviet Military Developments* (Dept. of the Army,

ACSI, Wash., D.C.: April 1977), one of a set of discussions on recent developments in the Soviet armed forces, is a good background source. Finally, *Brassey's Warsaw Pact Infantry and its Weapons* (J.I.H. Owen, Boulder, CO: Westview Press, 1976) describes ground forces and their weapons.

(c) Air Forces: Information on Soviet tactical air forces is contained in the "Soviet Aerospace Almanac" (see above); *Soviet Aerospace Handbook* (AFP 200-21) (DAF, Wash., D.C.), the Air Force counterpart to *Understanding Military Developments; The Observer's Soviet Aircraft Directory* (William Green and Gordon Swanborough, London: Frederick Warne, 1975), an often quoted source; and *Soviet Air Power in Transition* (Robert P. Berman, Wash., D.C.: Brookings, 1978). Other sources on Soviet air forces include the general military sources; *Soviet Aerospace,* a private newsletter; and articles appearing in the general military magazines.

(d) Naval Forces: The *Guide to the Soviet Navy* (Siegfried Breyer and Norman Polmar, Annapolis, MD: USNI, 1977) is the best compilation of information on the organization, ships and weapons, naval air and infantry forces, shipbuilding and bases and ports of the Soviet Navy. It also contains a good bibliography. An annotated bibliography of writing on Soviet naval forces and developments is *The Soviet Navy, 1941-1978: A Guide to Sources in English* (Myron J. Smith, comp., Santa Barbara, CA: ABC-Clio Press, 1980), which includes all relevant English language sources, a list of periodicals, and an index.

A number of excellent books and anthologies are also available on the Soviet Navy:

Soviet Naval Development: Capability and Context (Ken Booth, Michael MccGwire, and John McDonnell, eds., New York: Praeger, 1973).

Soviet Naval Developments: Objectives and Constraints (Ken Booth, Michael MccGwire, and John McDonnell, eds., New York: Praeger, 1975).

Soviet Naval Influence: Domestic and Foreign Dimensions (Michael MccGwire and John McDonnell, eds., New York: Praeger, 1977).

Soviet Naval Diplomacy (Bradford Dismukes and James McConnell, eds., New York, Pergamon Press, 1979).

Securing the Seas: The Soviet Naval Challenge and Western Alliance Options (Paul H. Nitze, Leonard Sullivan, Jr., and the Atlantic Council Working Group, Boulder, CO: Westview, 1979).

The Navy has also produced an introduction to the Soviet Navy, *Understanding Soviet Naval Developments* (3d Ed.) (DN, Wash., D.C.: GPO, 1977), which is a counterpart to the Army *Understanding Soviet Military Developments* and the Air Force *Soviet Aerospace Handbook*. It is updated by *Understanding the Soviet Navy: A Handbook* (Robert B. Bathhurst, Newport, RI: Naval War College Press, 1979). *Jane's Fighting Ships* is a required source on the Soviet naval forces. The other sources listed in Section VA and IIB are also essential. Soviet naval developments are reported regularly in the naval periodicals.

5. Soviet Military Doctrine and Strategy: Most of the sources on Soviet military doctrine and strategy are translations of Soviet writings by scholars and military writers and officers. The preeminent collection of works is the continuing Air Force series of translations, the *Soviet Military Thought* series, which now numbers 15 volumes. All of these volumes are available from the GPO and they cover various works of the Soviet military press. Table 34 is a listing of the titles in the series.

The classic volume on Soviet military strategy and doctrine is *Military Strategy* (Voennaya Strategiya) (Harriet Fast Scott, tr./ed., New York: Crane, Russak, 1975) by Soviet Marshal V.D. Sokolovski. This work appears as a number of different translations. *The Sea Power of the State* (Admiral of the Fleet of the Soviet Union Sergei G. Gorshkov, Annapolis, MD: Naval Institute Press, 1979) is a translation of the classic doctrinal work on the seapower of the Soviet Union. The Defense Intelligence Agency publishes a number of unclassified reports on Soviet doctrine and strategy which are worth reading. These studies are listed on Table 35.

Other sources on Soviet doctrine and strategy are traceable through the three bibliographic sources already mentioned: *Soviet Sources of Military Doctrine and Strategy*, "The Soviet Military Press, 1978," and "Soviet Military Doctrine in the 70s."

6. Soviet Military Spending: Beginning in 1976, the CIA began making available analyses of Soviet military spending. This series of CIA publications provides information on Soviet programs and spending and also fuels big controversy in the Defense analysis community on this issue:

Estimated Soviet Defense Spending in Rubles, 1970-75 (CIA, SR 76-10121U, May 76).

A Dollar Comparison of Soviet and U.S. Defense Activities, 1965-1975 (CIA, SR 76-10053, Feb. 76).

TABLE 34
SOVIET MILITARY THOUGHT SERIES

Vol. I	*The Offensive* (A.A. Sidorenko, 1979) (GPO, 1973)
Vol. II	*Marxism-Leninism on War and Army* (1972) (GPO, 1974)
Vol. III	*Scientific-Technical Progress and the Revolution in Military Affairs* (N.A. Lomov, ed., 1973) (GPO, 1974)
Vol. IV	*The Basic Principles of Operational Arts and Tactics* (V. Ye. Savkin, 1972) (GPO, 1974)
Vol. V	*The Philosophical Heritage of V.I. Lenin and Problems of Contemporary War* (A.S. Milovidov, ed., 1972) (GPO, 1974)
Vol. VI	*Concept, Algorithm, Decision* (V.V. Druzhinin and D.S. Kontorov, 1972) (GPO, 1975)
Vol. VII	*Military Pedagogy* (COL A.H. Danchenko and COL I.F. Vydrin, 1973) (GPO, 1976)
Vol. VIII	*Military Psychology* (V.V. Shelyag, et al., 1972), (GPO, 1976)
Vol. IX	*Dictionary of Basic Military Terms* (1965) (GPO, 1976)
Vol. X	*Civil Defense* (P.T. Yegorov, et al., 1970) (GPO, 1976)
Vol. XI	*Selected Soviet Military Writings* (GPO, 1977)
Vol. XII	*The Armed Forces of the Soviet State* (A.A. Grechko, 1975) (GPO, 1977)
Vol. XIII	*The Officer's Handbook* (S.N. Kozlov, ed., 1971) (GPO, 1977)
Vol. XIV	*The People, The Army, The Commander* (COL H.P. Skirdo, 1970)
Vol. XV	*Long-Range, Missile-Equipped: A Soviet View* (B.A. Vasilyev) (GPO, 1979)

A Dollar Cost Comparison of Soviet and U.S. Defense Activities, 1966-1976 (CIA, SR 77-10001U, Jan. 77).

A Dollar Cost Comparison of Soviet and U.S. Defense Activities, 1967-1977 (CIA, SR 78-10002, Jan. 78).

A Dollar Cost Comparison of Soviet and U.S. Defense Activities, 1968-1978 (CIA, SR 79-10004U, 1979).

Soviet and U.S. Defense Activities, 1970-1979: A Dollar Cost Comparison (CIA, SR 80-10005U, 1980).

Soviet and U.S. Defense Activities, 1971-1980: A Dollar Cost Comparison (CIA, SR 81-10005U, 1981).

Estimated Soviet Defense Spending Trends and Prospects (CIA, SR 78-10121, June 78).

TABLE 35
DIA PUBLICATIONS ON THE SOVIET UNION

Bibliography of Unclassified Books and Monographs on the Soviet and East European Ground Forces
Comparative Dictionary of U.S.-Soviet Terms (DDI-2200-33-77)
Medical Support of the Soviet Ground Forces (DDB-1150-18-79)
Physical Training of the Soviet Soldier (DDB-2680-48-78)
Soviet and Warsaw Pact Exercise: 1976 (DDI-1100-159-77)
Soviet/Warsaw Pact Ground Forces Camouflage and Concealment Techniques (DDI-1100-161-78)
Soviet and Warsaw Pact River Crossing: Doctrine and Capabilities (DDI-1150-13-77)
Soviet Ground Forces: Night Operations (DDI-1100-128-76)
The Soviet Ground Forces Training Program (DDB-1100-200-78)
Soviet Military Operations in Built-Up Areas (DDI-1100-155-77)
The Soviet Motorized Rifle Battalion (DDB-1100-77-76)
The Soviet Motorized Rifle Company (DDI-1100-77-76)
Soviet Tactics: The Meeting Engagement (DDI-1100-143-76)
Soviet Tank Battalion Tactics (DDI-1120-10-77)
Soviet Tank Company Tactics (DDI-1120-129-76)
Soviet Tank Regiment Tactics (DDB-1120-12-79)
Women in the Soviet Armed Forces (DDI-1100-109-76)

A former CIA analyst, William T. Lee, has written a great deal on the subject, criticizing CIA estimates as being too low. His major work is *The Estimation of Soviet Defense Expenditures, 1955-1975: An Unconventional Approach* (William T. Lee, New York: Praeger, 1977). There are a number of other works by this author on this subject.

These are by no means the only sources of information and data on Soviet spending. The *SIPRI Yearbook, The Military Balance,* and *World Military and Social Expenditures* also contain estimates. Numerous Congressional publications deal with the spending issue. The *Allocation of Resources in the Soviet Union and China* hearings of the Joint Economic Committee, and specific hearings by the House Intelligence and Armed Services Committees in 1980, examine estimates of Soviet spending and methodologies. A number of RAND Corporation reports have also been done. Finally, the "Counterpart of Defense Industry Conversion . . ." (see Section VC1), lists sources on the Soviet economy, military expenditures, research and development and the methodology for the measurement of military expenditure.

D. WESTERN EUROPE AND NATO

Much material is available on NATO and the European nations. Study of the Western European militaries in the context of NATO and the European military balance is quite popular and there is generally good raw and secondary material available. In fact, much more information is available on the European military systems and European military and strategic affairs than on any other region of the world. The many military magazines reporting procurement of systems, arms sales and military cooperation, force structure changes, exercises, readiness and capabilities, provide a wealth of material, almost equalling the amount of material available on the U.S. military.

1. Reference Tools: The only comprehensive bibliography of writing on NATO is *The Atlantic Alliance: A Bibliography* (Colin Gordon, New York: Nichols Publishing Co., 1978) which has many drawbacks being unannotated and retrospective from 1945-1977. Another bibliography on European military issues is the poorly-named *Nuclear Weapons and NATO: An Analytical Survey of Literature* (DOD, Wash., D.C.: GPO, 1975), which actually covers NATO, general Western European defense, the strategic and theater balance, and country military forces. *NATO: Selected References* (Maxwell AFB, AL: Air University Library, May 1979) is a much more up-to-date and comprehensive bibliography vis-a-vis individual countries. Three other reference tools are *The European Communities: A Guide to Information Sources* (J. Brian Collester, Detroit, MI: Gale Research, 1979), an annotated bibliography of over 1,400 sources on the political-military aspects of the European community; *Sources of Information on the European Communities* (Dorris M. Palmer, ed., London: Mansell, 1979), a guide to legal, technical, commercial, industrial and economic sources on the European Economic Community and the EEC countries; and *Foreign Affairs Research Special Papers Available: Europe and Canada* (Dept. of State, Wash., D.C.: GPO).

2. Current Issues and Events: Since Europe is a major focus of the media, much appears in the general press. There are also a number of specialized military publications in the English language. Some of these publications are:

ADIU Report
Allied Interdependence Newsletter
Atlantic Community Quarterly
Defence Today
Military Technology and Economics
NATO Review
NATO's Fifteen Nations
Scandinavian Review
Western European Review
Yugoslav Survey

NATO's Fifteen Nations has the best coverage of military and strategic affairs vis-a-vis NATO and Western Europe and a number of other general military magazines (see Sections VA and IIB) also provide good current coverage of European military developments. The two CIA services, FBIS and JPRS, also provide valuable current information on European developments. *FBIS Daily Report: Vol. II, Western Europe* (FBIS, Springfield, VA: NTIS) translates broadcasts and editorials; and the JPRS *Translations on Western Europe* (JPRS, Springfield, VA: NTIS) translates articles and items from the written press and specialized media.

3. The United Kingdom: Certainly, more material is available on Britain and British military forces and issues than on any other country. Because of the common language, and the keen interest in Britain in military affairs, much information exists on the British military.

There are a number of government sources on British military plans, programs and spending available from Her Majesty's Stationery Office (HMSO). The *Government Publications Catalogue* (London: HMSO, 1923-) is a monthly listing and cumulation with an author, title and subject index. Separate "Sectional Catalogues" are also available from the HSMO, many of them dealing with military and strategic affairs. List 8 covers publications of the Aeronautical Research Council; List 67 lists publications of the Ministry of Defence (MOD) and the Navy, Army and Air Force Department; and List 69 covers publications on Overseas Affairs.

The official document on British military posture and spending is the annual Defence White Paper, *Statement on the Defence Estimates, 19__* (London: HMSO), which describes current military programs and military spending. Other recurring publications of the HMSO dealing with the military include *Defence Accounts, 19__-__* (London: HMSO), a recounting of the actual defense expenditures at the end of the fiscal year; and *Defence Supply Estimates* (London: HMSO), an annual procurement and investment report.

Other information on the British military appears in a number of miscellaneous publications:

Britain, 19__: An Official Handbook (London: HMSO), an annual reference book including a chapter on current British defense policy, armed forces organization and military commitments.

The Royal Air Force Yearbook, 19__ (London: Ducimus Books, Ltd.), a review of RAF aircraft, squadrons and forces with some articles on RAF in NATO and the European military balance.

British Defence Equipment Guide, 19__ (Published by authority of the MOD (Defence Sales Organization) (Eton, Berkshire, U.K.: Whitton Press), an annual source on British military equipment and companies. This guide includes three volumes: Products; more Products and Index of Products; and Companies and Trade Names. This is a good source on the British military industrial complex.

The organization and policy of the armed services is discussed in the official directives of the Ministry of Defence. These publications provide background and policy information on many aspects of the military:

The Queen's Regulations for the Army (MOD, London: HMSO, 1961-), with amendments and explanation of terms.

Regulations for the Territorial and Army Volunteer Reserve (MOD, London: HMSO, 1967-), with amendments.

The Queen's Regulations and Air Council Instructions for the Royal Air Force (MOD, London: HMSO, 1956-), with amendments.

The Queen's Regulations of the Royal Navy (MOD, London: HMSO, 1967-), with amendments.

Background information on the British military and British military and strategic affairs appears in *British Defence Policy in a Changing World* (John Baylis, ed., London: Croom Helm, Ltd., 1977), a well-indexed and referenced anthology of articles covering policy and trends in Britain. *The United Kingdom's*

Current Defence Program and Budget (David Greenwood, Aberdeen: Univ. of Aberdeen, 1978), and the various *Aberdeen Studies in Defence Economics* series (King's College, Aberdeen: Univ. of Aberdeen), cover British defense programming, budgeting and spending alternatives.

The British military magazines are also an excellent source of information, containing data on deployments, forces and exercises, as well as doctrine and policy. The *ADIU Report* contains a regular bibliography of articles and publications on British defense issues, as well as news on Parliamentary developments. Table 36 contains a list of specialized British military periodicals.

4. Northern European Defense Issues: A good bibliography on Scandinavia and northern Europe is *Scandinavia: A Bibliographic Survey of Literature* (DA Pam 550-18) (DA, Army Library, Wash., D.C.: GPO, 1975), a selective and annotated bibliography of the Army Library series. Much has subsequently been written on Northern European defense and the military forces of the Northern European countries, much of which provides good background information. Some of the recommended works are:

TABLE 36
BRITISH MILITARY PERIODICALS

ADIU Report	Journal of the Royal Electrical
Arms Control and	and Mechanical Engineers
Disarmament	Mars and Minerva
Army Quarterly and	Navy News
Defence Journal	Pegasus
Aviation News	Provost Parade
Communicator	Recognition Journal
Defence Attache	Royal Air Force College Journal
Defence Materiel	RAF News
The Engineer	The RAOC Gazette
Globe and Laural	Royal Military Police Journal
Guards Magazine	The Royal Engineers Journal
Gunner	Soldier
The Hawk	Tank
Journal of the Royal	TAVR Magazine
Artillery	Waggoner

The Future of the Nordic Balance (Nils Andrem, Stockholm, Sweden: SSLP/Stockholm MOD, 1978), a study by the director of International Studies at the National Defense Research Institute, Stockholm.

The Defence Forces of Finland (Tavistock, Devon, U.K.: Army Quarterly and Defence Journal, 1974), a monograph on the organization and status.

Yearbook of Finnish Foreign Policy (Helsinki, Finland: Finnish Institute of International Affairs), an annual review.

The Defence Forces of Sweden (Tavistock, Devon, U.K.: Army Quarterly and Defence Journal, 1975), a monograph on the organization and status.

Defence Policy for the 70s and the 80s (MOD, Stockholm, Sweden, 1974).

The Evolution of Doctrines and the Economics of Defence (Swedish MOD, Stockholm, Sweden, 1973).

Documents on Swedish Foreign Policy (Ministry of Foreign Affairs, Stockholm, Sweden), an annual compilation.

A primary source of information on Northern European issues are the numerous research institutes in Scandinavia which conduct much of their research in the English language: the Stockholm International Peace Research Institute (SIPRI), Peace Research Institute, Oslo (PRIO), and Tampere Peace Research Institute.

5. *Central European Defense Issues:* Some basic sources are available on the nations of Central Europe, their military systems and the military situation in Central Europe.

A number of good English language sources exist on West Germany. The regular White Paper on Defense—*White Paper: The Security of the Federal Republic of Germany and the Development of the Federal Armed Forces*—is available from the Bundespresse and Informationsamt (53 Bonn, West Germany), and is issued about every 18 months.

Two other reports of the government available in English are *The Force Structure of the Federal Republic: Analysis and Options* (1973), a report by the Force Structure Commission (WehrstrukturKommission); and *Security and Defense: The Policy of the Federal Republic of Germany* (Ministry of Defense, Bonn, FRG, 1977). Another recent source on the Federal Republic is "Civil Military Relations in the Federal Republic of Germany," (Ralf Zoll, ed., in *Armed Forces and Society,* Summer 1979, entire issue).

An English language *White Paper on National Defence* is

published by France and is available from the Ambassade de France, Service de Presse et d'Informacion, 972 Fifth Avenue, New York, NY 10021. A 1973 *White Paper on French Nuclear Tests* is also available from the above address.

Two other sources on countries in Central Europe include:

The Defence Force of Switzerland (Tavistock, Devon, U.K.: Army Quarterly and Defence Journal, 1974), a monograph on organization and status.

The Defence Forces of Austria (Tavistock, Devon, U.K.: Army Quarterly and Defence Journal, 1975), a monograph on organization and status.

6. Southern European Defense Issues: Sources on southern Europe include writings on the southern European nations and the Mediterranean region. An English language *White Paper on Defense 1977: The Security of Italy and the Problems of the Italian Armed Forces* (Italian MOD, Rome, Italy, 1978) was issued in 1978 and gives good insight into Italian views of European defense and NATO. "The Revised Italian Defense System and the Reorganization of the Army" (*Italy: Documents and Notes,* Rome 26:99-110, 1979) discusses recent changes in the Italian military, and "The Italian Defense Forces and Their Problems," (Alberto Li Gobbi, in *RUSI and Brassey's Defence Yearbook 1978/79*), discusses organization and command and control.

Some other sources relating to strategic affairs in Southern Europe are:

Issues in U.S. Relations with Spain and Portugal (Report prepared for the Subcommittee on Europe and the Middle East, House Foreign Affairs Committee, by the CRS, 96-1, Wash., D.C.: GPO, 1979).

Turkey and U.S. Interests (CRS, (IB 79089), Wash., D.C.: 1979).

The Military Aspects of Banning Arms Aid to Turkey (Hearings before the Senate Armed Services Committee, 96-1, Wash., D.C.: GPO, 1978).

Turkey's Problems and Prospects: Implications for U.S. Interests (Report by CRS, LC, for the Subcommittee on Europe and the Middle East, House Foreign Affairs Committee, 96-2, Wash., D.C.: GPO, Mar. 3, 1980).

Turkey, Greece and NATO; The Strained Alliance (Staff Report to the Senate Foreign Relations Committee, 96-2, Wash., D.C.: GPO, Mar. 1980).

7. *NATO:* The NATO Information Service (1110 Brussels, Belgium) publishes a number of informative documents and handbooks on various aspects of NATO. These publications cover military, economic and scientific aspects of European defense and cooperation. *NATO Basic Documents* (1976) and the *NATO Handbook* (annual) provide background and historical information. Other sources of background information are the major commands of NATO which have press and information packages on their commands and interests. These commands are:

> Supreme Allied Commander, Atlantic
> Norfolk, VA 23511
>
> Supreme Allied Commander, Europe
> APO NY 09055
>
> HQ, Allied Forces, Northern Europe
> APO NY 09085
>
> HQ, Allied Forces, Central Europe
> APO NY 09011
>
> HQ, Allied Forces, Southern Europe
> FPO NY 09524

The NATO military magazine, *NATO Review*, is a valuable official resource, containing articles, documents and news items on NATO, Western European defense and the European military balance. Some Allied Regulations are unclassified and are also of value. *Allied Military Organization and Command* (AAP-1) (Designations-Short Titles-Locations-Charts) presents excellent information on NATO organization, including information on the location and designation of NATO commands, agencies and activities.

Information sources on the NATO forces have already been discussed in Sections VA and IIB and earlier in this section. DMS *Market Intelligence Reports: NATO Weapons* provides a wealth of material on NATO armed forces, plans, programs and activities. *NATO Infantry and its Weapons* (J.I.H. Owen, ed., Boulder, CO: Westview, 1976) provides information on ground forces and weapons. Another source, *From Guns to Butter: Technology Organizations and Reduced Spending in Western Europe* (Bernard Udis, Cambridge, MA: Ballinger, 1978), includes descriptions of the military industry in a number of Western European nations.

The posture of NATO is discussed as part of the hearings on

the Defense Budget annually. The Defense Department submits an annual report to the Congress on NATO, *Rationalization/ Standardization within NATO: Report, January 19__: A Report to the United States Congress by the Secretary of Defense* (Wash., D.C.: DOD, OASD (ISA)), which reviews U.S. participation and actions taken to improve rationalization, standardization, interoperability, and military cooperation. Another annual report, of the North Atlantic Assembly and published by the Senate Foreign Relations Commitee, is a source of information on the state of NATO military, political, and economic issues.

The Congressional Budget Office publishes a series of reports on NATO and U.S. forces in Europe which are of value. These reports provide a good amount of data and various policy alternatives relating to NATO forces and the conventional forces in Europe:

Strengthening NATO: POMCUS and Other Approaches (Feb. 1979).
U.S. Air and Ground Conventional Forces for NATO: Firepower Issues (Mar. 1978)
U.S. Air and Ground Conventional Forces for NATO: Air Defense Issues (Mar. 1978).
U.S. Air and Ground Conventional Forces for NATO: Mobility and Logistics Issues (Mar. 1978).
U.S. Air and Ground Conventional Forces for NATO: Overview (Jan. 1978).
Planning U.S. General Purposes Forces: The Theater Nuclear Forces (Jan. 1977).

Three Brookings studies dealing with U.S. forces in Europe and the European military balance are:

U.S. Troops in Europe: Issues, Costs and Choices (John Newhouse, et al., Wash., D.C.: Brookings, 1971).
U.S. Force Structure in NATO: An Alternative (Richard D. Lawrence and Jeffrey Record, Wash., D.C.: Brookings, 1974).
U.S. Nuclear Weapons in Europe: Issues and Alternatives (Jeffrey Record, Wash., D.C.: Brookings, 1974).

Views of the European military situation are presented in two reports by Sam Nunn, a conservative critic of the adequacy of U.S. and NATO forces against the Soviet forces. These two reports have been influential in the debate relating to the NATO military situation:

NATO and the New Soviet Threat (Report by Senator Sam Nunn and Senator Dewey Bartlett to the Senate Armed Services Committee, 95-1, Wash., D.C.: 1977), an evaluation of NATO forces and defense in light of the build-up of Soviet forces in Eastern Europe.

Policy, Troops and the NATO Alliance (Report of Senator Sam Nunn to the Senate Armed Services Committee, Wash., D.C.: GPO, 1974).

Other assessments of the European military situation are reports that have appeared in the various *Military Balances* and the *SIPRI Yearbooks*. An unbiased account of the problems involved in assessing the state of the military balance is another CBO study—*Assessing the NATO/Warsaw Pact Military Balance* (CBO, Wash., D.C.: GPO, Dec. 1977) which discusses the various methods of assessment and their drawbacks.

8. Canada: Information on the Canadian military and Canadian military and strategic affairs is generally good. The Canadian government publishes some reports on military issues and foreign affairs traceable through *Canadian Government Publications: Monthly Catalog* (Canadian Department of Public Printing and Stationary, Ottawa: The Queen's Printer, 1953-), which has an index and an annual cumulative issue. Two annual Department of National Defense reports are necessary reading: *Defence 19__* (Canadian Department of National Defense (DND), Ottawa, Canada: The Queen's Printer) and *Public Accounts of Canada* (Ottawa, Canada: The Queen's Printer). *Defence* is a review of the Armed Forces, their organization, operations and commitments. *Public Accounts* presents figures on military spending and arms sales and assistance. Other information is also available from the Information Service of the Canadian DND (Dept. of National Defence Information Service, Ottawa, Ont., K1A 0K2, Canada). *The Queen's Regulations and Orders for the Canadian Forces* (Ottawa, Ont., Canada: S&S Canada, Pub. Ctr., Printing & Publishing) is a loose-leaf service of the regulations and orders for the Canadian military which outlines policy and organization.

Information on Canadian foreign policy is available from *International Perspectives: A Journal of Opinion on World Affairs,* published by the Department of External Affairs of the Canadian government. This monthly journal reports on Canadian foreign policy and includes a reference section listing recent books and articles on Canadian foreign policy, publications of the Department, and recent treaties and agreements of the Canadian government. *International Canada: The Events of . . .* is a private monthly which includes information on Canadian foreign policy and reports international developments vis-a-vis Canada. *Canadian Foreign Relations* (Canadian Department of External Affairs, Information Division, Ottawa, Canada) is an annual review of the state of Canada's foreign relations. More information on Canadian foreign policy is obtainable from the Department of External Affairs, Public Relations, Ottawa, Ont., K1A 0G2, Canada. *The Canadian Annual Review of Politics and Public Affairs* (John Saywell, ed., Toronto, Ont./Buffalo, NY: Univ. of Toronto Press, 1960-) is a private review of Canadian government and politics with a section on external affairs.

Current information on Canada is available from *Canadian News Facts: The Indexed Digest of Canadian Current Events* (Toronto, Ont.: Harpep, 1967-), a biweekly in-depth coverage of Canadian news including external affairs and defense. A number of specialized Canadian military and international periodicals may also be of interest:

> *Air Force*
> *Army, Navy and Air Force Journal*
> *Canadian Defence Quarterly*
> *Canadian Military Engineer*
> *Canadian Military Journal*
> *Canadian Sailor*
> *International Canada*
> *International Journal*
> *International Perspectives*
> *Ploughshares Monitor*
> *Sentinel*

Sentinel is the official magazine published by the Canadian Department of National Defence Information Service and is an excellent general review of Canadian military affairs.

E. ARMS CONTROL AND DISARMAMENT

Sources of information on current arms control negotiations, theories and actions are widely produced and distributed and one of the most accessible categories for the researcher. A number of government documents and publications, Congressional publications, United Nations publications and other country and private organization publications all relate to the theory and practice of arms control. For the scope of this guide, those sources which have reference value or are recurring are listed.

1. Reference Tools: The volume of written material on arms control is great, the area being the favorite of academics and politicians for theoretical writing. Some bibliographies of this material are:

Arms Control and Disarmament: A Bibliography (Richard D. Burns, comp., Santa Barbara, CA: ABC-Clio Press, 1977), the best and most comprehensive bibliography of almost 9,000 titles.

Arms Control, Disarmament and Economic Planning: A List of Sources (Judith Roswell, Los Angeles: Center for the Study of Armament and Disarmament, CSU, 1973), a small (300 titles) bibliography.

Disarmament: A Select Bibliography, 1973-1977 (New York: United Nations, 1978), the latest edition of a bibliography that is updated at five-year intervals by the Dag Hammarskjold Library of the United Nations.

Peace Research: Definitions and Objectives: A Bibliography (Kinde Durkee, Los Angeles, CA: Center for the Study of Armament and Disarmament, CSU, 1976) a small bibliography.

SALT II: A Bibliography (Julia F. Carlson and Robert G. Bell, Library of Congress, CRS (Report 78-176F), 1 Sept. 78), an annotated list of over 200 titles covering the January 1972-August 1978 period.

The SALT Era: A Selected Bibliography (Revised edition) (Richard Dean Burns and S.H. Hutson, Los Angeles, CA: Center for the Study of Armament and Disarmament, 1979), a small bibliography.

SALT II: A Selected Research Bibliography (Jerry Parker and Thomas A. Meeker, Los Angeles, CA: Center for the Study of Armament and Disarmament, CSU, 1973), a small bibliography.

Selected SALT Bibliography (Dennis G. Wills, New York: Columbia Univ. Press, 1977).

NPT: Current Issues in Nuclear Proliferation: A Selected Bibliography (Susan Ridgeway, comp., Los Angeles, CA: Center for the Study of Armament and Disarmament, CSU, 1977), a small bibliography.

Bibliography: Nuclear Proliferation (Report for the Committee on International Relations and the Committee on Science and Technology of the House by the CRS, LC, Wash., D.C.: GPO,

April 1978), an annotated bibliography of 2,500 sources.

The Proliferation of Nuclear Weapons and the Nonproliferation Treaty (NPT) (A Selective Bibliography and Source List) (Thomas A. Meeker, Los Angeles, CA: Center for the Study of Armament and Disamament, CSU, 1973), a small bibliography.

2. Background: General information sources on arms control from an international perspective are definitely lacking. Some of the United Nations reports fill the gap, but generally deal only with U.N. activities and developments, and not the political context of arms control. The *United Nations Disarmament Yearbook* (New York: Centre for Disarmament, 1977-) reviews U.N. activities, including the texts of major treaties and conventions. It is, however, normally too late for current research. *Arms Control: A Survey and Appraisal of Multilateral Agreements* (Stockholm, Sweden: SIPRI, 1978) was prepared as a reference source for the U.N. Special Session on Disarmament and still contains analysis of the major existing treaties and agreements (including the texts) for reference.

A number of sources on U.S. policy and agreements also provide background information. *Arms Control and Disarmament Agreements: Texts and History of Negotiations* (ACDA, Wash., D.C.: GPO, 1977-), a biennial publication, includes a discussion of every major arms control agreement since 1925 to which the United States is a party, with the text of the agreement. *Documents on Disarmament* (ACDA, Wash., D.C.: GPO, 1960-) is an annual compilation of documents relating to arms control. The *Arms Control Report: Annual Report to Congress* (ACDA, Wash., D.C.: GPO, 1962-) discusses current policy and developments in arms control and disarmament and the status of negotiations. *The Fiscal Year 19__ Arms Control Impact Statements* (Joint Committee Print for use of the Committee on Foreign Relations and Foreign Affairs of the Senate and the House, Wash., D.C.: GPO, 1978-) is a compilation of statements prepared by the Arms Control and Disarmament Agency (ACDA) describing current weapons systems and R&D programs, their military necessity and their impact on U.S. arms control policy including their effects on treaties, negotiations, stability, forces and verification. An analysis of the Arms Control Impact Statements (ACIS) prepared by the CRS and submitted in connection with the budget is *Evaluation of Fiscal Year 19__ Arms Control Impact Statements: Towards More Informed Congressional Participation in National Security Policymaking* (Report by the CRS as Joint Committee Print for the Committees of Foreign Relations and Foreign Affairs, U.S.

House, U.S. Senate, Wash., D.C.: GPO, 1978-).

Growth in technology is important to arms control. A number of works relate to the issue of technology and its relationship to arms control:

Outer Space: Battlefield of the Future, (SIPRI, New York: Crane, Russak, 1978), a description of the basic concepts of satellites and the military uses of space (with glossary and references).

Anti-Personnel Weapons (SIPRI, New York: Crane, Russak, 1979), a discussion of the relationship between military utility and humanity and arms control.

Nuclear Proliferation Factbook (Joint Committee Print for the House Foreign Affairs Committee/Senate Foreign Relations Committee, prepared by the CRS, LC, Wash., DC: GPO, 1980), basic documents, data and statistics on nuclear proliferation and nuclear weapons and technology with many valuable tables and charts, a glossary and bibliography.

U.S. Foreign Policy on Spent Nuclear Fuel (CRS, (IB 79107), Wash., D.C.: 1979).

Chemical Weapons and Chemical Arms Control (Matthew Messelson, ed., Wash., D.C.: Carnegie Endowment for International Peace, 1978), a well-documented reference work on chemical weapons with an appendix describing U.S. and NATO weapons and sources of information on Soviet and Warsaw Pact weapons.

Chemical/Biological Warfare: A Selected Bibliography (Julian P. Robinson, comp., Los Angeles, CA: Center for the Study of Armament and Disarmament, CSU, 1979), a small bibliography.

Herbicides as Weapons: A Bibliography (Arthur H. Westing, Los Angeles, CA: Center for the Study of Armament and Disarmament, CSU, 1973).

Some periodicals dealing with arms control which provide background information and current news are:

ADIU Report
Alternatives
Arms Control
Arms Control and Disarmament
Arms Control Today
Bulletin of the Atomic Scientists
Defense Monitor
Disarmament

Disarmament Campaigns
Disarmament Times
FAS Newsletter/Public Interest Report
FCNL Washington Newsletter
Orbis
Survival

A number of peace research journals are also valuable sources for research in the field of arms control:

Bulletin of Peace Proposals
Chronicle
Cooperation and Conflict
Current Research on Peace and Violence
Journal of Conflict Resolution
Journal of Peace Research
Journal of Peace Science
Peace and Change
Peace and the Sciences
Peace Research
Peace Research Reviews

F. INTERNATIONAL ORGANIZATIONS AND INTERNATIONAL LAW

Inclusion of a section on International Law and Organization is more of an aside than a complete analysis of the sources available in this specialized area. There are some general sources, however, that the researcher may need to use, and they are referenced here. A further discussion of the sources for International Law appears in Volume III of *The Information Sources of Political Science* by Holler. This should be consulted for more in-depth study.

1. International Law: The writing on international law as it relates to military and strategic affairs is traceable through a couple of bibliographies and one abstracts and index service. *Public International Law: A Current Bibliography of Articles* is a semiannual unannotated bibliography of works on international law, including war and armed conflict, international organizations, and sea, air and space law. *The International Bibliography of Air Law: 1900-1971* (Wybo P. Heere, Dobbs

Ferry, NY: Oceana Publications, 1972) includes sources on air law including military aviation and reconnaissance. *Law and Politics in Outer Space: A Bibliography* (Irving L. White, et al., Tucson, AZ: Univ. of Arizona Press, 1972) is a retrospective bibliography listing sources dealing with the political and legal problems in space.

Sources for current developments in international law affecting military and strategic affairs are *The American Journal of International Law* and *Ocean Development and International Law.* The *American Journal* is the major American journal in international law and has an extensive book review and bibliographic section which reviews new books and announces the publication of documents and agreements.

The State Department submits an annual report—*Digest of United States Practice in International Law* (Dept. of State, Office of the Legal Advisor, Wash., D.C.: GPO, 1973-)—which reports developments and practice vis-a-vis the United States.

Information on treaties and alliances is available from a few standard reference books. In Section IVG1, sources for treaties and alliances of the United States were already discussed. Sources for international information include the following:

World Treaty Index and *Treaty Profiles* (Peter H. Rohn, ed., Santa Barbara, CA: ABC-Clio Press, 1974), a comprehensive reference book, well-indexed and referenced, which analyzes world treaty patterns using statistical information and comparative statistics.

Treaties and Alliances of the World (3d Ed.) (Bristol, U.K./New York: Keesing's Publications/Scribner's, 1974) provides the basic information about alliances throughout the world and the treaties and agreements creating those alliances. The texts of the agreements are cited in the pertinent issues of *Keesing's Contemporary Archives,* which also serves as the current reference work on the subject of signings and abrogations.

United Nations Treaty Series (New York: United Nations, 1946-) includes the texts of all treaties and other international agreements registered with the United Nations, with a chronological, subject and country index.

2. *International Organizations:* Information on the many international organizations is available in a number of yearbooks which include data on international and regional bodies. The *International Yearbook and Statesmen's Who's Who, The Europe Yearbook* and *The Statesmen's Yearbook* all include sections dealing with international organizations and agencies. The *Yearbook of International Organizations* (Brussels, Belgium: Union of International Associations, 1948-) describes more than 4,000 governmental and non-governmental organizations (including sources of further information) and is supplemented monthly by the periodical *International Organizations*. The *Yearbook of International Congress Proceedings* (Brussels, Belgium: Union of International Associations, 1969-) reports developments and actions of the various organizations and agencies.

3. The United Nations: The United Nations is a great producer of paper on various aspects of international law, its activities, crisis and negotiations, and disarmament. United Nations material is indexed in *The UNDEX: United Nations Documents Index* (New York: United Nations, 1950-), which is issued monthly (with annual cumulative index) and indexes by subject, country and document publications of the various agencies of the U.N. Three research guides which describe in detail the U.N. publications system and the various sources of information are:

Publications of the United Nations System: A Reference Guide (Harry N.M. Winton, ed., New York: Bowker, 1972), a comprehensive guide to the U.N. and its publications including a list of periodicals of the U.N. system. This guidebook is kept up to date by entries in *International Bibliography: Information and Documentation,* a quarterly publication of Bowker.

Guide to U.N. Organizations, Documentation and Publishing for Students, Researchers, Librarians (Peter L. Hajnal, Dobbs Ferry, NY: Oceana, 1978) describes the structure and functions, publications and bibliography of sources on the U.N.

A Guide to the Use of U.N. Documents, Including Reference to the Specialized Agencies and Special U.N. Bodies (Brenda Brimmer, Dobbs Ferry, NY: Oceana, 1962), a description of the documents classification system and various approaches to research on U.N. issues.

Two annual reviews of the United Nations and its activities are *Annual Review of the United Nations* (Dobbs Ferry, NY: Oceana, 1949-) and *Yearbook of the U.N.* (New York: U.N., Dept. of Public Information, 1947-). U.S. participation in the United Nations is outlined in two publications: *U.S. Participation in the U.N.: Report by the President to the Congress* (Dept. of State, Wash., D.C.: GPO, 1947-), an annual report, and *U.S. Contributions to International Organizations, Annual Report* (Dept. of State, Wash., D.C.: GPO).

APPENDIX A: MILITARY AND STRATEGIC SERIALS AND PERIODICALS

Throughout this guide, the value of periodicals has been stressed because of the raw data and the current information they report. Under each subject area a list of pertinent periodicals has been mentioned. Table 37 lists the areas of classification and the pages in this guide where a listing appears. A number of

TABLE 37
SUBJECT LISTING OF PERIODICALS

Africa	154
Air Forces	143
Armed Forces	26, 89
Arms Control	193
Arms Sales	131
Armies	93, 145
Asia	158
Australia	158
Bibliographic Resources (General)	6
Canada	190
Communications and Electronics	150
D.O.D. (Published by Dept. of Defense)	52, 56, 57, 60
Ground Forces	145
India	160
International Relations	25, 26
Latin America	167
Middle East	169
Military Associations (pro-military)	109
Military Sciences	26
Navies	60, 147
Newsletters and Information Services	89, 138
Peace Research	194
Personnel Policy and Issues	114
Regional Defense Issues	154, 158, 167, 169, 173, 182
Soviet Union	173
Strategic Issues	26
U.K. Military	184
U.S. Air Force	56

U.S. Armed Forces	52, 56, 57, 60
U.S. Army	57
U.S. Defense Posture	26, 52, 89
U.S. Navy	60
U.S. Programs, Spending and Contracting	26, 52, 89, 109
U.S. Reserves	97
Weapons	138, 143, 145, 147
Western Europe	182

---

monographs and serials dealing with military and strategic affairs are of great interest. Table 38 lists the major monographs and serials and the organizations which publish them.

The 400 military and strategic related periodicals chosen for inclusion in this appendix are generally well known among analysts in each sub-field. For each periodical, information is given on frequency of publication (D—Daily, W—Weekly, BW—Bi-weekly, M—Monthly, BM—Bi-monthly, Q—Quarterly, SA—Semiannually), length of publication (if available), and indexing in one of the three major specialized military indexes (AU—Air University Index, NA—Naval Abstracts, AMB—Abstracts of Military Bibliography). Those periodicals marked with an * are publications of the U.S. government, available on subscription from the Government Printing Office.

TABLE 38
SERIALS AND MONOGRAPHS IN MILITARY AND STRATEGIC AFFAIRS

Aberdeen Studies in Defense Economics, Centre for Defence Studies, Univ. of Aberdeen
Adelphi Papers, International Institute for Strategic Studies
Agenda Papers, National Strategy Information Center
American-Asian Educational Exchange Monograph Series, Institute of Far Eastern Studies, Seton Hall University
Arms Control and International Security Working Papers, Center for International and Strategic Affairs, UCLA
Atlantic Papers, The Atlantic Institute for International Affairs
Atlantic Council Security Series Policy Papers, Atlantic Council of the United States
Behind the Headlines Series, Canadian Institute of International Affairs, Toronto
Canberra Papers, The Australian National University Center for Strategic and Defence Studies

Center for International Policy Reports, Center for International Policy

Conflict Studies, Institute for the Study of Conflict

Editorial Research Reports, Congressional Quarterly

Foreign Policy Research Institute Monographs, FPRI

Foreign Policy Reports (and Special Reports), Institute for Foreign Policy Analysis

Harvard Studies in International Affairs, Harvard University

Headline Series, Foreign Policy Association

IPS Issue Papers and Books, Institute for Policy Studies

Jerusalem Papers on Peace Problems, Hebrew University

Mershon Center Position Papers, Mershon Center, Ohio State University

Military Issues Research Memorandum, Strategic Studies Institute, U.S. Army War College

Monographs in International Affairs Series, Advanced International Studies Institute

Monograph Series, Lehigh University

Monograph Series in World Affairs, University of Denver

National Security Affairs Monographs, National Defense University

National Strategy Information Center Monograph Series, NSIC

Occasional Papers, Armament and Disarmament Information Unit, University of Sussex

Occasional Papers, Aspen Institute for Humanistic Studies

Occasional Papers, Center for the Study of Armament and Disarmament, California State University

Occasional Papers, Cornell University Peace Studies Program

Occasional Papers, Stanley Foundation

Occasional Papers in International Affairs, Center for International Affairs, Harvard University

Peace Research Reviews, Canadian Peace Research Institute

Policy Papers in International Affairs, Institute for International Studies, University of California, Berkeley

SAGE Professional Papers in International Studies, SAGE

Seaford House Papers

Special Reports, U.S. Strategic Institute

Studies in International Affairs Series, Institute of International Studies, Columbia University

Studies in Defense Policy, Brookings Institution

Strategic and Defence Studies Centre (SDSC) Working Papers

Strategy Papers, National Strategy Information Center

Vantage Conference Reports, Stanley Foundation

War, Revolution and Peacekeeping, SAGE

The Washington Papers, CSIS, Georgetown University

Working Papers, International Security Studies Program, The Wilson Center

Working Papers, Institute for World Order

PERIODICALS ON MILITARY AND STRATEGIC AFFAIRS

Accounting and Finance Tech Digest
Air Force Accounting and Finance Center
Denver, CO 80279
(BW), 1950-

ADIU Report
Armament and Disarmament Information Unit
Mantell Building
University of Sussex, Falmer, Brighton BN1 9RF, U.K.
(BM), 1979-

Advisor
Office of Civilian Manpower Mgmt.
DN (Code 58)
Washington, D.C. 20390
(Q)

The Advocate
USA Legal Service Agency
HQDA (JAAJ—DD), Nassif Bldg.
Falls Church, VA 22041
(BM), 1969-

AEI Foreign Policy and Defense Review
American Enterprise Institute for Public Policy Research
1150 17th St., N.W.
Washington, D.C. 20036
(M/10), AU, NA

Aeronautical Quarterly
Royal Aeronautical Society
4 Hamilton Place
London W1V OBQ U.K.
(Q), 1949-, NA

Aerospace
Royal Aeronautical Society
4 Hamilton Place
London W1V OBQ, U.K.
(M/10), NA

Aerospace Daily
Ziff-Davis Aviation Division
1156 15th Street, N.W.
Washington, D.C. 20005
(D)

Aerospace Historian
Dept. of History
Eisenhower Hall
Kansas State University
Manhattan, KS 66506
(Q), 1954-, AU, AMB

Aerospace Intelligence
Defense Marketing Service, Inc.
100 Northfield Street
Greenwich, CT 06830
(W)

Aerospace International
Monch Publishing Group
HeilbachStrasse 26
5300 Bonn 1, FRG
(BM), AU, NA, AMB

*Aerospace Safety Magazine
AFISC/SEDEP
Norton AFB, CA 92409
(M), AU

Africa (An Independent Business Economic and Political Monthly)
Africa Journal Ltd.
Kirkman House, 54A
Tottenham Court Road
London W1P 0BT, U.K.
(M)

Africa Currents
Africa Publishing Trust
48 Grafton Way
London W1P 5LB, U.K.
(Q)

Africa Diary
Africa Publications
F-15 Bhagat Singh Market,
Box 702
New Delhi, India
(W), 1961-

Africa News
P.O. Box 3851
Durham, NC 27702
(W)

Africa Report
African American Institute
833 United Nations Plaza
New York, NY 10017
(BM)

Africa Today
  (A Quarterly Review)
Africa Today Associates
Graduate School of International
  Studies
University of Denver
Denver, CO 80208
(Q)

Africa Affairs
  (The Journal of
  the Royal African Society)
Centre of International and Area
  Studies
15 Woburn Square
London WC 1 HONS, U.K.
(Q), NA

Africana Journal
Africana Publishing Co.
30 Irving Place
New York, NY 10003
(Q)

Air Combat
Challenge Publications, Inc.
7950 Deering Avenue
Canoga Park, CA 91304
(BM)

Air Defense Magazine
USA Air Defense School
Ft. Bliss, TX 79916
(Q), AU, NA, AMB

The Air Force Administrator
HQ, USAF (AS/DASJ)
Washington, D.C. 20330
(M)

*Air Force Comptroller
HQ, USAF (AF/AC)
Washington, D.C. 20330
(Q), AU

*Air Force Engineering and
  Services Quarterly*
HQ, USAF/LEE
Washington, D.C. 20330
(Q), AU

Air Force Law Review
AF Jag School
Maxwell AFB, AL 36112
(Q), AU

Air Force Magazine
Air Force Association
1750 Pennsylvania Avenue, N.W.
Washington, D.C. 20006
(M), 1918-, AU, NA, AMB

*Air Force Policy Letter for
  Commanders*
HQ, USAF (SAF/PA)
Washington, D.C. 20330
(BW)

*Air Force Policy Letter
  Supplement*
AFSINC/IIA
Kelly AFB, TX 78241
(M), AU

Air Force Times
Army Times Publishing Co.
475 School Street, S.W.
Washington, D.C. 20024
(W), 1940-, AU

Air International
P.O. Box 16
Bromley BR2 7RB, Kent, U.K.
(M), NA

Air Pictorial
Profile Books, Ltd.
Dial House
6 Park Street
Windsor SL4 1UU, Berks, U.K.
(M), NA

*Air Reservist
Bolling AFB
Washington, D.C. 20332
(M/10), 1949-, AU

**Air University Review*
Maxwell AFB, AL 36112
(BM), 1947-, AU, NA, AMB

AirForce (The Royal Canadian Air Force Association Magazine)
RCAF Association
424 Metcalfe St.
Ottawa, K2P 203, Ontario, Canada
(Q)

*Airman (Official Magazine of the USAF)
AFSINC
Kelly AFB, TX 78241
(M), 1957-, AU, NA

*All Hands (The Official Naval Magazine)
NIRA (NAVOP-0071)
Washington, D.C. 20350
(M), 1914-, NA

All Volunteer (The Army's Recruiting and Retention Professional Magazine since 1919)
U.S. Army Recruiting Command
Attn: USARCCS-PA
Ft. Sheridan, IL 60037
(M)

Allied Interdependence Newsletter
CSIS, Georgetown University
1800 K Street, N.W.
Washington, D.C. 20006
(M)

Alternatives (A Journal of World Policy)
Institute for World Order
777 United Nations Plaza
New York, NY 10017
(Q), NA

American Intelligence Journal (The Magazine for Intelligence Professionals)
National Military Intelligence Association, Inc.
1606 Laurel Lane
Annapolis, MD 21401
(Q), 1977-

American Journal of International Law
American Society of International Law
2223 Massachusetts Ave., N.W.
Washington, D.C. 20009
(Q)

AMPO: Japan-Asia Quarterly Review
Pacific-Asia Resources Center
P.O. Box 5250
Tokyo International, Japan
(Q), 1968-

An Cosantoir (The Irish Defense Journal)
Army Headquarters, Parkgate
Dublin 8, Ireland
(M), AMB

*Approach (The Naval Aviation Safety Review)
Naval Safety Center
NAS Norfolk, VA 23511
(M), NA

Arabia and the Gulf (The Independent Weekly Review of Arab Political and Economic Affairs)
Portico Pubs, Ltd.
84 Fetter Lane
London EC4A 1EQ, U.K.
(W), 1977-

Armada International
Weinbergstrasse 102
CH-8035, Zurich, Switzerland
(BM), NA

Armed Forces
Allied Publishing Ltd.
Laab Street, New Center
Johannesburg, RSA
(M)

Armed Forces and Society
  (An Interdisciplinary Journal)
Inter-University Seminar on
  Armed Forces and Society
Social Sciences Building
University of Chicago
1126 E. 59th Street
Chicago, IL 60637
(Q), 1974-, AU, NA, AMB

Armed Forces Comptroller
  (The Quarterly Professional
  Journal of the American Society
  of Military Comptrollers)
P.O. Box 91
Mt. Vernon, VA 22121
(Q), 1954-, AU

Armed Forces Journal
  International
Army & Navy Journal, Inc.
1414 22nd Street, N.W., Suite 603
Washington, D.C. 20037
(M), 1863-, AU, NA, AMB

Armor
  (The Magazine of Mobile Warfare)
U.S. Armor Association
P.O. Box O
Ft. Knox, KY 40121
(BM), AU, AMB

Arms Control
Frank Cass, Ltd.
Gainsborough House
Gainsborough Road
London E11 1R5, U.K.
(Q), 1980-

Arms Control and Disarmament:
  Developments in the International
  Negotiations
Arms Control and Disarmament
  Research Unit
Foreign and Commonwealth
  Office
Downing Street (East)
London SW1A 2AH, U.K.
(Q) 1979-

Arms Control Today
Arms Control Association
11 Dupont Circle, N.W.
Washington, D.C. 20036
(M), NA

Army
Association of the U.S. Army
2425 Wilson Blvd.
Arlington, VA 22201
(M), 1950-, AU, AMB

Army Administrator
  (The Magazine for Military
  Managers)
USA Admin. Center
Attn: ATZI-PAO
Ft. Benjamin Harrison, IN 46216
(BM)

Army Aviation
  (A Professional Journal
  Endorsed by the Army Aviation
  Assn. of America)
Army Aviation Publication, Inc.
1 Crestwood Road
Westport, CT 06880
(M/10)

The Army Communicator
  (Voice of the Signal Corps)
USA Signal Center
Signal Towers
Ft. Gordon, GA 30905
(Q), 1976-

*The Army Lawyer
The JAG School
Charlottesville, VA 22901
(M)

Army Logistician
(The Official Magazine of U.S. Army Logistics)
USA Logistics Management Center
Ft. Lee, VA 23801
(BM), AU, AMB

Army, Navy & Air Force Journal
Army, Navy and Air Force Veterans Association of Canada
Scarborough, Ontario, Canada
(Q), 1972-

Army Quarterly and Defense Journal
1 West Street
Tavistock, Devon, U.K.
(Q), 1829-, AU, NA, AMB

Army Research, Development & Acquisition
USA DARCOM, Attn: DRCDE-LN
5001 Eisenhower Avenue
Alexandria, VA 22333
(BM), NA

Army Reserve Magazine
Chief, Army Reserve
Attn: DAAR-PA
Washington, D.C. 20310
(Q)

Army Times
Army Times Publishing Co.
475 School Street, S.W.
Washington, D.C. 20024
(W), 1939-

Asia Monitor
Asia/North America Communications Center
2 Man Wan Road, 17-C
Kowloon, Hong Kong
(Q), 1976-

Asia Pacific Community
(A Quarterly Review)
Asian Club, Suite 2302
World Trade Center Bldg.
2-4-1 Hamamatsu-Cho
Minato-Ku, Tokyo, Japan
(Q), NA

Asia-Pacific Defense Forum
CINCPAC, Box 13
Camp H.M. Smith, HI 96861
(Q)

Asia Quarterly
(A Journal from Europe)
Centre d'Etude du Sud-Est Asiatique et de l'Extreme-Orient
Avenue Jeanna 44
B-1050 Brussels, Belgium
(Q), NA

Asia Research Bulletin
(Monthly Economic Reports with Political Supplement)
Asia Research Pte., Ltd.
Alexandra P.O. Box 91
Singapore 9115
(M)

Asian Affairs
Royal Society for Asian Affairs
42 Devonshire Street
London, W1, U.K.
(TA), 1903-, NA

Asian Affairs
(An American Review)
American-Asian Educational Exchange, Inc.
82 Morningside Drive
New York, NY 10027
(BM), 1973-, NA

Asian Defence Journal
Syed Hussain Publications (Sdn)
BMD, Penthouse, 6th Floor
Bangunan Bakti, Box 838
91 Jln Campbell
Kuala Lumpur, Malaysia
(BM), 1971-, NA

Asian Perspective
  (A Biennial Journal of
  Regional & International
  Affairs)
Institute for Far Eastern Studies
Kyungnam University
28-42 Samchung-Dong
Chongro-Ku, Seoul, Korea
(BA)

Asian Recorder (Weekly Record
  of Asian Events with Index)
H.S.R. Khemchand
C-2 Gulmohar Park, Box 595
New Delhi 110049, India
(W), 1955-

Asian Survey
Institute of International Studies
University of California
2234 Piedmont Avenue
Berkeley, CA 94720
(M), NA

Asiaweek
  (The Asian News Weekly)
Federal Building
359 Lockhart Road
Hong Kong
(W), 1975-

Assegai
  (The Magazine of the Rhodesian
  Army)
City Printers and Stationers Pty. Ltd.
Box 1943
Salisbury, Zimbabwe-Rhodesia
(M), 1961-

Assembly
Association of Graduates, USMA
West Point, NY 10996
(Q)

Astronautics & Aeronautics
  (A publication of the American
  Institute of Aeronautics and
  Astronautics)
1290 Avenue of the Americas
New York, NY 10019
(M/11), NA

Atlantic Community Quarterly
Atlantic Council of the U.S.
1616 H St., N.W.
Washington, D.C. 20006
(Q), NA

Aussen Politik
  (German Foreign Affairs
  Review) (English edition)
Interpress
Uebersee Verlag GMBH
Schoene Aussicht 23
D-2000 Hamburg 76, FRG
(Q), NA

Australian Foreign Affairs Record
Department of Foreign Affairs
Canberra, ACT 2600, Australia
(M)

The Australian Journal of Defense Studies
Business Mgr, DOD
Russell Offices
Canberra, ACT 2600, Australia
(SA), 1977-

The Australian Journal of Politics and History
Department of History
University of Queensland
St. Lucia 4067, Queensland
Australia
(TA), NA

Australian Outlook
  (Journal of the Australian
  Institute of Int'l Affairs)
Box E181, Post Office
Canberra, ACT 2600, Australia
(TA), 1947-, NA

Aviation & Marine International
  (The Int'l Technical, Economic
  and Political Magazine on
  Aviation and Naval Matters)
Sorecom S.A.M./Interinfo
Dept. 16, Rue des Orchidees
Monte-Carlo, Monaco
(M/11), NA, AMB

Aviation News (Britain's
   International Aviation
   Newspaper)
Alan W. Hall (publications), Ltd.
26 The Broadway
Amersham, HP7 OAR, Bucks, UK
(BW)

*Aviation, Space and
   Environmental Medicine*
Aerospace Medical Association
Washington National Airport
Washington, D.C. 20001
(M), 1930-, AU

*Aviation Week & Space
   Technology*
1221 Avenue of the Americas
New York, NY 10020
(W), 1916-, NA

Born in Battle
Eshel-Dramit Ltd.
P.O. Box 115
Hod Hasharon, Israel
(Q)

*British Journal of International
   Studies*
Longman Group, Ltd.
Periodicals and Directories Div.
Burnt Hill, Harlow
Essex CM20 2JE, U.K.
(TA), NA

Bulletin of Peace Proposals
International Peace Research
   Institute
Radhusgaten 4
Oslo 1, Norway
(Q), 1970-, NA

*The Bulletin of the Atomic
   Scientists*
   (A magazine of Science and
   Public Affairs)
Education Foundation for Nuclear
   Scientists
1020-24 E. 58th Street
Chicago, IL 60637
(M/10), 1945-, NA

*Campus
   (The Navy Education and
   Training Monthly)
Chief of Naval Education and
   Training Support
NAS Pensacola, FL 32508
(M)

*Canadian Aeronautics and Space
   Journal*
Canadian Aeronautics and Space
   Institute, Saxe Building
60-75 Sparks Street
Ottawa K1P 5A5, Canada
(Q), NA

Canadian Defence Quarterly
Defence Publications
100 Adelaide Street West
Toronto, Ontario MSH 153
Canada
(Q), NA, AMB

Canadian Military Engineer
National Defense HQ/DCMEO
Ottawa, Ontario K1A OK2
Canada
(SA), 1960-, AMB

The Canadian Military Journal
   (Independent Defence Forces
   Review)
3450 Durocher, Suite 8
Montreal PQ H2X 281, Canada
(Q), 1934-

Canadian Sailor
Seafarers Int'l Union of Canada
634 St. James Street
W. Montreal, Quebec, Canada
(M), 1950-

Caribbean Review
Florida International University
Tamiami Trail
Miami, FL 33139
(Q)

CEP Newsletter
Council on Economic Priorities
84 Fifth Avenue
New York, NY 10011
(M/10-12)

China Quarterly
Contemporary China Institute
School of Oriental and African
 Studies
Malet Street
London WC1, U.K.
(Q), 1960-, NA

China Report
Centre for the Study of Developing
 Societies
39 Rajpur Road
Delhi 110054, India
(BM)

Chronicle
The Hammarskjold Information
 Centre on the Study of Violence
 and Peace
68 Eton Place
Eton College Road
London NW3 2DS, U.K.
(Q), 1979-

CIC Update
Conversion Information Center
Council on Economic Priorities
84 Fifth Avenue
New York, NY 10011
(Q)

*Combat Crew
 (Magazine of the SAC)
HQ, SAC (IGFC)
Offutt AFB, NE 68113
(M), AU

Commanders Call
HQDA (SAPA-CI)
OCPA
Washington, D.C. 20310
(BM)

Communicator
Royal Navy
Communications Branch
Whitehall, London SW1, U.K.
(TA), 1947—

Comparative Political Studies
SAGE Publications, Inc.
275 S. Beverly Drive
Beverly Hills, CA 90212
(Q), 1968-, NA

Comparative Politics
Transaction Periodicals
 Consortium
Rutgers University
New Brunswick, NJ 08903
(Q), 1968-, NA

Comparative Strategy
 (An International Journal)
Crane, Russak and Co., Inc.
347 Madison Avenue
New York, NY 10017
(Q), NA, AMB

Conflict
 (An International Journal)
Crane, Russak and Co., Inc.
347 Madison Avenue
New York, NY 10017
(Q), NA

Conflict Studies
Institute for the Study of Conflict
12/12A Golde Square
London W1R 3AF, U.K.
(M), NA

Contemporary China
 (Developmental, Comparative
 and Global Perspectives)
East Asian Institute
Columbia University
420 West 118th St.
New York, NY 10027
(Q), NA

Continental Marine and Digest
JPAO/4th MarDiv/4th MAW/
 MCR
4400 Dauphine Street
New Orleans, LA 70146
(M)

Contracting and Acquisition Newsletter
HQ, USAF/RDCX
Washington, D.C. 20330
(M)

The Conversion Planner
(A Newsletter of Action on Economic Conversion)
SANE, A Citizens' Organization for a Sane World
514 C Street, N.E.
Washington, D.C. 20002
(BM)

Cooperation and Conflict
(Nordic Journal of International Politics)
Institute of Political Sciences
University of Aarhus
DK—8000 Aarhus C, Denmark
(Q), NA

Covert Action Information Bulletin
P.O. Box 50272
Washington, D.C. 20004
(BM)

Cuban Studies
Center for Latin American Studies
University of Pittsburgh
G6-Hervis Hall
Pittsburgh, PA 15260
(SA), NA

Current Digest of the Soviet Press
American Institute for the Advancement of Slavic Studies
1314 Kinnear Road
The Ohio State University
Columbus, OH 43212
(W), 1929-

Current History
(A World Affairs Monthly)
4225 Main Street
Philadelphia, PA 19127
(M/10), 1914-, NA

Current Research on Peace and Violence
Tampere Peace Research Institute
Hameekatu 13 b A
P.O. Box 447
SF-33101 Tampere-10, Finland
(Q)

Defence
Whitton Press Ltd.
50 High Street
Eton, Berks, UK
(M), 1970-, NA, AMB

Defence Africa
Tom Chalmers Enterprises, Ltd.
Granville House
St. Peters Street
Winchester, Hampshire, U.K.
(Q), 1974-

Defence Attache
Diplomatist Associates, Ltd.
Shooter's Lodge
Windsor Forest, Berkshire, U.K.
(Q), 1972-

Defence Force Journal
(A Journal of the Australian Profession of Arms)
Building C, Room 4-25
Russell Offices
Canberra, ACT 2600, Australia
(BM), 1976-, NA, AMB

Defence Journal
16-B 7th Central Street
Defence Housing Society
Karachi 4, Pakistan
(M), 1975-

Defence Management
Institute of Defence Management
Bolarum PO
Secunderabad 500010, India
(SA)

Defence Materiel
  (The Journal for the Worldwide
  Promotion of British Defence
  Equipment)
Eldon Publications, Ltd.
30 Fleet Street
London EC4Y 1AH, U.K.
(BM), 1976-, NA, AMB

Defence Science Journal
  (Journal of the R&D
  Organization)
Scientific Information
  and Documentation Centre
Metcalfe House
New Delhi, 110054, India
(Q), 1949-, AMB

Defence Today
Public & Consult International
Via Tagliamento No. 29
00198 Rome, Italy
(BM), NA

Defence/81
American Forces Press Service
1117 N. 19th Street
Arlington, VA 22209
(M), AU, NA

*Defense and Economy World
  Report and Survey*
Government Business Worldwide
  Reports
P.O. Box 4875
Washington, D.C. 20008
(W), 1969-

Defense & Foreign Affairs Daily
Copley & Associates, S.A.
2030 M Street, N.W., Suite 602
Washington, D.C. 20036
(D)

Defense & Foreign Affairs Digest
Copley & Associates, S.A.
2030 M Street, N.W., Suite 602
Washington, D.C. 20036
(M), AU, NA

Defense Daily
Space Pubs, Inc.
1341 G Street, N.W.
Washington, D.C. 20005
(D)

Defense Electronics
  (*Including Electronic Warfare*)
EW Communications, Inc.
1170 East Meadow Dr.
Palo Alto, CA 94303
(M), AU, NA, AMB

Defense Latin America
Tom Chalmers Enterprises, Ltd.
Granville House
St. Peters Street
Winchester, Hampshire, U.K.
(Q), 1976-

Defense Management Journal
OASD(MRAL)
Cameron Station
Alexandria, VA 22134
(BM), AU, NA, AMB

Defense Monitor
Center for Defense Information
122 Maryland Avenue, N.E.
Washington, D.C. 20002
(M/10), 1972-, AU, NA, AMB

*Defense Systems Management
  Review*
Defense Systems Management
  College
Ft. Belvoir, VA 22060
(Q), 1977-, AU, NA

Defense Transportation Journal
National Defense Transportation
  Assn.
1612 K Street, N.W.
Washington, D.C. 20006
(BM), AU, NA

Defense Week
300 National Press Building
Washington, D.C. 20045
(W), 1980-

Department of State Bulletin
(The Official Monthly Record of U.S. Foreign Policy)
Bureau of Public Affairs
Department of State
Washington, D.C. 20520
(M), 1939-, NA

Department of State Newsletter
DGP/PA, Room 3237
Department of State
Washington, D.C. 20520
(M/11)

Direction
(The Navy Public Affairs Magazine)
NIRA, Print Media Division
Hoffman #2, 200 Stovall St.
Alexandria, VA 22332
(BM)

DISAM Newsletter
Defense Institute of Security Assistance Management
Wright Patterson AFB, OH 45433
(Q), 1978—

Disarmament
(A Periodic Review by the U.N.)
Centre for Disarmament
United Nations
New York, NY 10017
(Q), 1978-

Disarmament Campaigns
(International Newsletter on Actions Against the Arms Race)
Nonviolent Alternatives
Kerkstraat 150
B-2000 Antwerp, Belgium
(Q), 1980-

Disarmament Times
777 United Nations Plaza
New York, NY 10017
(BM/8)

Driver
(The Traffic Safety Magazine for the Military Driver)
AFISC/SEDD
Norton AFB, CA 92409
(M)

The Engineer
U.S. Army Engineer School
Ft. Belvoir, VA 22060
(Q)

EurArmy Magazine
HQ, AFN-Europe
APO NY 09757
(M)

European Scientific News
Office of Naval Research
Branch Office, Box 39, London
FPO NY 09510
(M)

European Studies Newsletter
Council for European Studies
156 Mervis Hall
University of Pittsburgh
Pittsburgh, PA 15260
(BM), 1972-

Exchange and Commissary News
Executive Business Media, Inc.
P.O. Box 1500, 825 Old Country Road
Westbury, NY 11590
(M)

Far Eastern Economic Review
GPO, BOX 160
Hong Kong, BCC
(W), 1946-, NA

FAS Newsletter/Public Interest Report
Federation of American Scientists
307 Massachusetts Avenue, N.E.
Washington, D.C. 20002
(M)

Fathom
 (The Surface Ship and
 Submarine Safety Review)
Naval Safety Center
Norfolk, VA 23511
(Q), NA

FCNL Washington Newsletter
Friends Committee on National
 Legislation
245 Second Street, N.E.
Washington, D.C. 20002
(M)

Field Artillery Journal
Box 33131
Ft. Sill, OK 73503
(BM), AU, AMB

The Fletcher Forum
 (A Journal of Studies in
 International Affairs)
Fletcher School of Law
 and Diplomacy
Tufts University
Medford, MA 02155
(SA)

Flight International
IPC Transport Press, Ltd.
Dorset House
Stamford Street
London SE1 9LU, U.K.
(W), 1909-, NA

Foreign Affairs
Council on Foreign Relations
58 E. 68th Street
New York, NY 10021
(Q), 1922-, NA, AMB

Foreign Affairs Record
Ministry of External Affairs
External Publicity Division
New Delhi, India
(M)

Foreign Policy
345 E. 46th Street
New York, NY 10017
(Q), 1971-, NA

Foreign Service Journal
American Foreign Service Assn.
2101 E Street, N.W.
Washington, D.C. 20037
(M), 1924-

Free China Review
Chung Hwa Information Service
P.O. Box 337
Taipei, Taiwan
(M), NA

Globe and Laural
 (The Journal of the Royal
 Marines)
Royal Marines Eastney
Southsea, P04 9PX, Hampshire,
 U.K.
(BM), 1892-, AMB

*Government Contractors
 Communique*
Federal Publications Inc.
1725 K Street, N.W.
Washington, D.C. 20006
(BW)

Ground Defence International
Sorecom S.A.M./Interinfo
Dept. 16, Rue des Orchidees
Monte Carlo, Monaco
(M), AMB

Guards Magazine
Horse Guards
Whitehall, London SW1A 2AX,
 U.K.
(Q), 1892-

Gunner
 (Magazine for the Royal
 Artillery)
Royal Regiment of Artillery
Government House, New Road
Woolrich, London SE18, U.K.
(M), 1970-

The Hawk
 (The Independent Journal
 of the RAF Staff College)
RAF Staff College
Bracknell, RG12 30D, Berks, U.K.
(A)

Helicopter
Avia Press Associates
Delta House, Summer Lane
Worle, Weston-Super-Mare
Avon BS22 OBE, U.K.
(BM)

History, Numbers and War
 (A HERO Journal)
Historical Evaluation and
 Research Organization
P.O. Box 157
Dunn Loring, VA 22027
(Q), 1977-, NA

Hovering Craft and Hydrofoil
Kalerghi Publications
51 Welbeck Street
London, WIM 7HE, U.K.
(Q), 1961-, NA

IDSA Journal
 (Quarterly Journal of the
 Institute for Defense Studies
 and Analyses)
Sapru House
Barakhamba Road
New Delhi, 110001, India
(Q), AMB

India Quarterly
 (A Journal of International
 Affairs)
Indian Council of World Affairs
Sapru House
Barakhamba Road
New Delhi, 110001, India
(Q), 1945-, NA, AMB

Infantry
 (A Professional Journal for
 the Combined Arms Team)
U.S. Army Infantry School
Ft. Benning, GA 31905
(BM), AU, AMB

Intelligence Digest World Report
 (A Review of World Political,
 Economic and Strategic Affairs)
Intelligence International, Ltd.
17 Rodney Road
Cheltenham, GL50 IJQ
Gloucestershire, U.K.
(M)

Inter-American Economic Affairs
P.O. Box 181
Washington, D.C. 20044
(Q), 1947—

Interavia
 (World Review of Aviation,
 Astronautics, Avionics)
Interavia S.A.
86 Avenue Louis-Casai
P.O. Box 162
1216 Cointrin, Geneva
Switzerland
(M), 1946-, AU, NA

Interavia Air Letter
Interavia S.A.
86 Avenue Louis-Casai
CH-1216, Cointrin, Geneva
Switzerland
(D), 1933-

International Affairs
Polish Institute of Int'l Affairs
"Ars Polona"
Krakowski Przedmiescie 7
Warsaw, Poland
(M), 1963-

International Affairs
 (A Monthly Journal of
 Political Analysis)
All-Union Znaniye Society
14 Gorokhovsky Perulok
Moscow K-64, USSR
(M), 1955-, NA

International Affairs
(Journal of the Royal Institute of Int'l Affairs)
Royal Institute of Int'l Affairs
Chatham House, St. James Square
London SW1, U.K.
(Q), 1931-, NA, AMB

International and Comparative Law Quarterly
British Institute of Int'l and Comparative Law
32 Furnival Street
London, EC4A 1JN, U.K.
(Q), 1952-

International Canada
Canadian Institute of Int'l Affairs
15 King's College Circle
Toronto, M5S 2V9, Canada
(M)

International Defense Business
Government Business Worldwide Reports
P.O. Box 4875
Washington, D.C. 20008
(W)

International Defense Intelligence
Defense Marketing Service, Inc.
100 Northfield Street
Greenwich, CT 06830
(W)

International Defense Review
Interavia S.A.
86 Avenue Louis-Casai
CH-1216 Cointrin, Geneva
Switzerland
(M), AU, NA, AMB

International Interactions
(A Transnational Multi-disciplinary Journal)
University of North Carolina
Chapel Hill, NC 27514
(Q), 1974-, NA

International Journal
Canadian Institute of Int'l Affairs
230 Bloor Street, W.
Toronto 5, Ontario Canada
(Q), 1946-, NA, AMB

International Organization
(A Journal of Political and Economic Affairs)
200 Lou Henry Hoover Building
Stanford University
Stanford, CA 94305
(Q), 1947-, NA

International Perspectives
(A Journal of Opinion on World Affairs)
Department of External Affairs
Ottawa, K1A 059, Ontario, Canada
(BM), NA, AMB

International Policy Report
Center for International Policy
120 Maryland Avenue, N.E.
Washington, D.C. 20002
(M)

International Problems
Israeli Institute for the Study of International Affairs
2 Pinkser Street (P.O. Box 17027)
Tel Aviv, 61170, Israel
(Q), 1963-

International Relations
David Davies Memorial Institute of International Studies
Thorney House, Smith Square
London, SW1, U.K.
(BM), 1954-, NA

International Security
Program for Science and International Affairs,
Harvard University
73 Boylston Street
Cambridge, MA 02138
(Q), AU, NA, AMB

International Security Review
Center for International Security
  Studies
The American Security Council
  Education Foundation
Boston, VA 22713
(Q)

International Studies
  (Quarterly Journal of the School
  of Int'l Studies Jawaharlal
  Nehru University)
New Mehrauli Road
New Delhi, 110057, India
(Q)

International Studies Quarterly
SAGE Publications, Inc.
275 S. Beverly Drive
Beverly Hills, CA 90212
(Q), 1967-, NA

Issue
  (A Quarterly Journal of
  Africanist Opinion)
African Studies Association
Epstein Service Building
Brandeis University
Waltham, MA 02154
(Q)

The JAG Journal
OJAG, Department of the Navy
Washington, D.C. 20370
(SA), NA

Jane's Defence Review
Paulton House
8 Shepherdess Walk
London N1 7LW, U.K.
(BM), 1980-

Japan Quarterly
Asahi Shimbun
Yuraku-Cho
Chiyoda-Ku, Tokyo, Japan
(Q), 1953-

*Jerusalem Journal of
  International Relations*
The Leonard Davis Institute
  for International Relations
Hebrew University
Givat Ram
Jerusalem, Israel
(Q), 1974-

Joint Perspectives
Armed Forces Staff College
Norfolk, VA 23611
(Q), 1980-

Journal of Aircraft
  (Devoted to Aeronautical
  Science and Technology)
American Institute of Aeronautics
  and Astronautics
1296 Avenue of the Americas
New York, NY 10019
(Q)

Journal of Asian Studies
Association for Asian Studies
University of Washington
Seattle, WA 98195
(Q), NA

Journal of Conflict Resolution
Political Science Department
Yale University
124 Prospect Street
New Haven, CT 06520
(Q), 1957-, NA, AMB

Journal of Contemporary Asia
Box 49010
Stockholm 49, Sweden
(Q), 1970-, NA

Journal of Electronic Defense
Horizon-House-Microwave, Inc.
610 Washington Street
Dedham, MA 02026
(BM)

Journal of International Affairs
School of International Affairs
Columbia University
New York, NY 10027
(SA), 1947-, NA

Journal of Modern Africa Studies
  (A Quarterly Survey of Politics, Economics and Related Topics in Contemporary Africa)
University of Malawi
P.O. Box 278
Zomba, Malawi
(Q)

Journal of Palestine Studies
  (A Quarterly on Palestinian Affairs and the Arab-Israeli Conflict)
The Institute for Palestine Studies and Kuwait University
P.O. Box 19449
Washington, D.C. 20036
(Q), 1971-

Journal of Peace Research
International Peace Research Institute, Oslo
Radhusgaten 4
Oslo 1, Norway
(Q), 1964-, NA

Journal of Peace Science
  (An International Journal of the Scientific Study of Conflict and Conflict Management)
Peace Science Society
Department of Peace Science
University of Pennsylvania
3718 Locust Street
Philadelphia, PA 19174
(SA), 1973-, NA

Journal of Political and Military Sociology
Department of Sociology
Northern Illinois University
Dekalb, IL 60115
(SA), 1973-, AU, NA, AMB

Journal of Southeast Asian Studies
University of Singapore
Jalan Boon Lay
Jurong 22, Singapore
(SA), 1970-

The Journal of Strategic Studies
Frank Cass & Co., Ltd.
Gainsborough House
Gainsborough Road
London E11 1RS, U.K.
(TA), 1978-

The Journal of the Army Intelligence and Security Command
HQ, INSCOM
Arlington, VA 22212
(M)

Journal of the Royal Artillery
The Regimental Secretary
Royal Artillery Institution
Woolwich, London SE1B 4JJ, U.K.
(SA), 1858-

Journal of the Royal Electrical and Mechanical Engineers
REME Institution
Aldershot, U.K.
(Q)

Korea and World Affairs
Research Center for Peace and Unification
C.P.O. Box 6545
Seoul, Korea
(Q), NA

Korea Observer
Academy of Korean Studies
Yonsei University
Box 3410
Seoul, Korea
(Q), 1968-

Latin American Digest
Center for Latin American Studies
Arizona State University
Tempe, AZ 85281
(Q), 1965-

Latin American Political Report
90-93 Cowcross Street
London EC1N 6BL, U.K.
(W), 1966-

Latin American Research Review
The Latin American Studies Assn.
3166 Hamilton Hall
University of North Carolina
Chapel Hill, NC 27514
(TA)

Leatherneck
Marine Corps Association
Box 1775
Quantico, VA 22134
(M)

The Log
Bancroft Hall
U.S. Naval Academy
Annapolis, MD 21412
(M)

*The Mac Flyer
Scott AFB, IL 62225
(M), AU

Maintenance
AFISC/SEDEP
Norton AFB, CA 92409
(Q)

Marine Corps Gazette
Marine Corps Association
Box 1775
Quantico, VA 22134
(M), AU, NA, AMB

Maritime Defence
(The Journal of International Naval Technology)
Eldon Publications Ltd.
30 Fleet Street
London EC4Y 1AH, UK
(M), NA, AMB

Mars and Minerva
Special Air Services Regiment
Combined Service Publications, Ltd.
London, UK
(M)

*Mech
(The Naval Aviation Maintenance Safety Review)
Naval Safety Center
Norfolk, VA 23511
(Q)

Medical Service Digest
HQ, USAF/SGI
Bolling AFB,
Washington, D.C. 20332
(M)

MEED Arab Report
Meed House
21 John Street
London WC1N 2BP, UK
(BW0, 1979-

MERIP Reports
Middle East Research and Information Project, Inc.
1470 Irving St., N.W.
Washington, D.C. 20010
(M/9)

Middle East Economic Digest
(Weekly News, Analysis and Forecast)
Meed House
21 John Street
London WC1N 2BP, UK
(W), 1957-

Middle East International
21 Collingsham Road
London SW5 ONU, UK
(BW), NA

The Middle East Journal
Middle East Institute
1761 H Street, N.W.
Washington, D.C. 20036
(Q), NA

Milavnews
(Monthly Newsletter Supplement to the International Air Forces and Military Aircraft Directory)
Aviation Advisory Services, Ltd.
Stapleford Airfield
Romford, Essex, U.K.
(M), 1964-

Military Affairs
(The Journal of Military History, Including Theory, and Technology)
Eisenhower Hall
Kansas State University
Manhattan, KS 66506
(Q), 1937-, NA, AMB

Military Chaplain
Military Chaplains Assn. of the U.S.
7758 Wisconsin Ave., Suite 401
Bethesda, MD 20014
(BM)

Military Digest
Director of Military Training
General Staff Branch MT4
Army Headquarters, DHQ, PO
New Delhi, India
(Q), 1973-

Military Digest
Military Training Directorate
General Staff Branch
General Headquarters
Rawalpindi, Pakistan
(Q), 1950-

Military Electronics/ Countermeasures
Hamilton Burr Pub. Co., Inc.
2065 Martin Ave., Suite 104
Santa Clara, CA 95050
(M), AU, NA

The Military Engineer
(Journal of the Society of American Military Engineers)
800 17th Street, N.W.
Washington, D.C. 20006
(BM), 1920-, AU, NA, AMB

Military Intelligence
P.O. Box 2603, USAICS
Ft. Huachuca, AZ 85613
(Q), AMB

Military Law Review
The JAG School, USA
Charlottesville, VA 22901
(Q)

Military Market Magazine
(Magazine for the Military Retail System)
Army Times Publishing Co.
475 School Street, S.W.
Washington, D.C. 20024
(M)

Military Media Review
DINFOS/PAO
Ft. Benjamin Harrison, IN 46216
(Q)

Military Medicine
Assn. of Military Surgeons of the U.S.
P.O. Box 104
Kensington, MD 20795
(M)

Military Police Law Enforcement Journal
USAMPS/TC
Ft. McClellan, AL 36205
(Q), 1973-

Military Research Letter
Callahan Publications
6631 Old Dominion Drive
McLean, VA 22101
(BW), 1958-

Military Review
USA Command & General
 Staff College
Ft. Leavenworth, KS 66027
(M), 1922-, AU, NA, AMB

*Military Technology and
 Electronics*
Wehr und Wissen
Hinsbachstrasse 26
P.O. Box 110-187
Bonn-Duisdorf, FRG
(BM), 1978-

Millenium
 (Journal of International
 Studies)
London School of Economics
Houghton Street
London, WC2A 2AE, U.K.
(TA)

Missiles/Ordnance Letter
Callahan Publications
6631 Old Dominion Drive
McLean, VA 22101
(BW), 1957-

Modern China
SAGE Publications, Inc.
275 S. Beverly Drive
Beverly Hills, CA 90212
(Q)

Multinational Monitor
P.O. Box 19312
Washington, D.C. 20036
(M)

Munitions Control Newsletter
Dept. of State
Office of Munitions Control
Washington, D.C. 20520
(M)

NACLA Report on the Americas
North American Congress
 on Latin America
151 West 19th Street, 9th Floor
New York, NY 10011
(M/10)

National Defense
American Defense Preparedness
 Assn.
740 15th Street, N.W.
Washington, D.C. 20005
(BM), AU, NA, AMB

National Guard
National Guard Assn. of the U.S.
1 Massachusetts Ave., N.W.
Washington, D.C. 20002
(M/11), AU

National Security Record
 (A Report on the Congress
 and National Security Affairs)
Heritage Foundation
513 C Street, N.E.
Washington, D.C. 20002
(M)

National Security Review
National Defense College of
 the Philippines
Ft. Bonafacio
Rizal, Philippines
(Q), 1973-

NATO Review
NATO Information Service
1110 Brussels, Belgium
(BM), NA, AMB

NATO's Fifteen Nations
 (Independent Review of
 Economic, Political and
 Military Power)
Jules Perel's Publishing Co.
P.O. Box 913
35 Matterhorn
Amstelveen 1186EB
Netherlands
(BM), AU, NA, AMB

Naval Affairs
Fleet Reserve Association
1303 New Hampshire Ave., N.W.
Washington, D.C. 20036
(M)

*Naval Aviation News
Bldg. 146, WNY
Washington, D.C. 20374
(M), NA

Naval Engineers Journal
The American Society of Naval
    Engineers, Inc.
1012 14th St., N.W., Suite 807
Washington, D.C. 20005
(BM), NA, AMB

Naval Forces
    (International Forum for
    Maritime Power)
Monch UK, Ltd.
31 A High Street
Aldershot
Hants GU 11 1BN, U.K.
(Q), 1980-

Naval Record
Pearce Pubs., Ltd.
1-3 Bank Street
Tonbridge, Kent TN9 1BL, U.K.
(BM), NA

*Naval Research Logistics
    Quarterly
Office of Naval Research
800 N. Quincy St.
Arlington, VA 22217
(Q), NA

*Naval Research Reviews
Office of Naval Research
Code 733
Arlington, VA 22217
(M), NA

Naval Reservist News
Chief of Naval Reserve
4400 Dauphine Street
New Orleans, LA 70146
(M)

Naval War College Review
USN War College
Newport, RI 02844
(Q), 1948-, AU, NA, AMB

The Navigator
323d FTW/DOTN
Mather AFB, CA 95655
(M)

*Navy Civil Engineer
Naval School/CEC Officers
Port Hueneme, CA 93043
(Q)

The Navy Human Resources
    Journal
NAS Memphis
Millington, TN 38054
(Q)

Navy International
    (An Independent Journal
    that Fosters Understanding of
    the Vital Importance of Sea
    Power)
72 High Street
Haslemere, Surrey GU27 21A, U.K.
(M), NA, AMB

*Navy Lifeline
    (The Navy Safety Journal)
Naval Safety Center
NAS Norfolk, VA 23511
(BM), NA

Navy News
Royal Navy
Whitehall, London SW1, U.K.
(M), 1953-

Navy Policy Briefs
NIRA/OCHINFO
Room 2E329, Pentagon
Washington, D.C. 20350
(M)

Navy Supply Corps Newsletter
    (The Professional Journal
    of the Navy Supply Corps)
NAVSUP (09D2)
Washington, D.C. 20376
(M), NA

Navy Times
Army Times Publishing Co.
475 School Street, S.W.
Washington, D.C. 20024
(W), 1950-, NA

NCOA Journal
P.O. Box 2268
San Antonio, TX 78298
(M)

Near East Report
   (Washington Letter on
   American Policy in the Middle
   East)
444 N. Capitol Street, N.W.
Washington, D.C. 20001
(W)

New African
I.C. Publications, Ltd.
Room 1121
122 E. 42nd Street
New York, NY 10017
(M), 1966-

The Officer
Reserve Officer Assn. of the U.S.
1 Constitution Avenue
Washington, D.C. 20002
(M), AU

Orbis
   (A Journal of World Affairs)
Foreign Policy Research Institute
3508 Market Street, Suite 350
Philadelphia, PA 19104
(Q), 1957-, NA, AMB

Pacific Affairs
   (An International Review of
   Asia and the Pacific)
University of British Columbia
2075 Wesbrook Mall
Vancouver, BC, V6T 1W5, Canada
(Q), NA

Pacific Defence Reporter
P.O. Box 235
Mona Vale, 2103 Australia
(M), 1978-, NA, AMB

Pacific Research
Pacific Studies Center
867 W. Dana St., #204
Mountain View, CA 94041
(Q)

Pakistan Army Journal
Pakistan MOD
General HQ
Rawalpindi, Pakistan
(SA)

Pakistan Horizon
The Pakistan Institute
   of International Affairs
Karachi-1
Aiwan-E-Sadar Road, Pakistan
(Q)

Parameters
   (Journal of the U.S. Army
   War College)
USAWC
Carlisle Barracks, PA 17013
(Q), AU, NA, AMB

Paratus
   (Official Magazine of the
   S.A. Defence Force)
Northvaal Building
Vermeulen Street
Pretoria 0002, South Africa
(M)

Peace and Change
   (A Journal of Peace Research)
Sonoma State University
Rohnert Park, CA 94928
(Q), 1972-

Peace and the Sciences
International Institute for Peace
Mollwaldplatz 5
A-1040 Vienna, Austria
(Q), 1979-

Peace Research
(A Monthly Journal of Original Research on the Problem of War)
Canadian Peace Research Institute
119 Thomas Street
Oakville, Ontario, Canada
(Q), 1969-

Peace Research Reviews
25 Dundana Ave.
Dundas, Ontario, L9H 4E5
Canada
(BM)

Pegasus
(Journal of the Parachute Regiment and Airborne Forces)
Browning Barracks
Aldershot GU11 2B5, Hants
U.K.
(Q)

Pioneer
(Singapore Armed Forces News)
Pers Res & Ed Dept.
Manpower Division, MOD
Tanglin, Singapore 10
(M), 1969-

Ploughshares Monitor
"Project Ploughshares"
Institute of Peace and Conflict Studies
Conrad Grebel College
University of Waterloo
Waterloo, Ontario N2L 3G6
Canada
(BM)

*Problems of Communism
International Communications Agency
1776 Pennsylvania Avenue, N.W.
Washington, D.C. 20547
(BM), NA

Program Manager
Defense Systems Management College
Ft. Belvoir, VA 22060
(BM), 1972-

Provost Parade
RAF Police School
RAF Newton
Nottingham, NG13 8HR, UK
(SA), 1947-

RAF News
(The Official Newspaper of the RAF)
Turnstile House
94-99 High Holborn
London, WC1V 6LL, UK
(BW), 1961-

The RAOC Gazette
(The Journal of the Royal Army Ordnance Corps and Army Ordnance Services)
RAOC Secretariat
Deepcut, Camberley, Surrey, UK
(M)

Renegotation/Management Letter
Callahan Publications
6631 Old Dominion Drive
McLean, VA 22101
(BW)

Resource Management Journal
Office of the Comptroller, DA
Washington, D.C. 20310
(Q), 1980-

The Retired Officer
(Published for the Seven Uniformed Services)
Retired Officers Association
1625 Eye St., N.W.
Washington, D.C. 20006
(M)

The Review
American Logistics Association
5205 Leesburg Pike, #1213
Falls Church, VA 22041
(BM), AU

Review of International Affairs
P.O. Box 413
11001 Belgrade, Yugoslavia
(BW), 1949-

The Round Table
  (The Commonwealth Journal
  of International Affairs)
18 Northumberland
London, WC2N 5AP, U.K.
(Q), 1910-, NA

Royal Air Force Journal
RAF College
Cranwell, Sleaford
Lincs, U.K.
(A), 1920-

The Royal Engineers Journal
The Institute of Royal Engineers
Chatham, Kent ME4 4UG, U.K.
(Q), 1870-

Royal Military Police Journal
Roussillon Barracks
Chicester, Sussex, U.K.
(Q), 1950-

RUSI Journal
  (Journal of the Royal United
  Service Institution for
  Defence Studies)
Royal United Service Institute
Whitehall, London SW1A 2ET
U.K.
(Q), AU, NA, AMB

The Russian Report
  (Inside Coverage of Soviet
  Affairs and Military
  Technologies)
Intelligence Press Service
A Division of Industry Reports,
  Inc.
7620 Little River Turnpike
Suite 414
Annandale, VA 22003
(M), 1980-

Russian Review
  (An American Quarterly
  Devoted to Russia Past and
  Present)
Russian Review, Inc.
Stanford, CA 94305
(Q), 1941-, NA

SAM
  (Soldier, Sailor, Airman,
  Marine)
1117 N. 19th St.
Arlington, VA 22209
(M), 1979-

Sane World
SANE
514 C Street, N.E.
Washington, D.C. 20002

Sanity
Campaign for Nuclear
  Disarmament
14 Gray's Inn Road
London WC1, U.K.

Scandinavian Review
American Scandinavian
  Foundation
127 E. 73rd Street
New York, NY 10021
(Q), 1913-

Scientific American
415 Madison Avenue
New York, NY 10017
(M)

Sea Technology
Compass Publications, Inc.
1117 N. 19th Street, Suite 1000
Arlington, VA 22209
(M), NA, AMB

*Sealift
  (Magazine of the Military
  Sealift Command)
MSC, Department of the Navy
4228 Wisconsin Avenue, N.W.
Washington, D.C. 20016
(M), NA

Seapower
(The Official Publication of the Navy League of the U.S.)
Navy League of the U.S.
818 18th Street, N.W.
Washington, D.C. 20006
(M), AU, NA, AMB

Security Police Digest
AFOSP/PA
Kirtland AFB, NM 87117
(M)

Sentinel
Department of National Defense
Information Services
Ottawa, Ontario, Canada
(BM), NA, AMB

Shield Newsletter
Whitten Press
50 High Street
Eton, Berkshire, U.K.
(BW)

Ships Monthly
Waterway Productions, Ltd.
Kottingham House
Dale Street
Burton-on-Trent, DE14 3TD, U.K.
(M), NA

Signal
(Journal of the Armed Forces Communications and Electronics Association)
AFCEA
5205 Leesburg Pike
Falls Church, VA 22041
(M/10), AU, NA, AMB

Slavic Review
(American Quarterly of Soviet and East European Studies)
American Assn. for the Advancement of Slavic Studies
409 E. Chalmers St., Room 359
University of Illinois
Champaign, IL 61820
(Q), 1941-

Soldier
(British Army Magazine)
Ministry of Defence
Ordnance Road
Aldershot, GU11 2DU, Hants, U.K.
(M)

Soldier of Fortune
(The Journal of Professional Adventurers)
Omega Group, Ltd.
5735 Arapahoe Avenue
Boulder, CO 80303
(M), AMB

Soldiers
(The Official U.S. Army Magazine)
Alexandria, VA 22314
(M), AU

South Africa International
South Africa Foundation
888 17th Street, N.W.
Washington, D.C. 20006
(Q), NA

South African Journal of African Studies
Africa Institute
Box 630
Pretoria, RSA
(A)

Southeast Asia Chronicle
S.E. Asia Resource Center
P.O. Box 4000-D
Berkeley, CA 94704
(BM)

Southern Africa
17 West 17th Street
New York, NY 10011
(M)

Soviet Aerospace
Space Publications, Inc.
1341 G Street, N.W.
Washington, D.C. 20005
(W)

Soviet Analyst
   (A Fortnightly Newsletter)
P.O. Box 39
Richmond, Surrey, U.K.
(BW)

Soviet Military Review
2 Marshall Biryuzon Street
Moscow 123298, USSR
(M), AU, NA

Soviet Press: Selected
   Translations
AFIS/INC
Directorate of Soviet Affairs
Bolling AFB
Washington, D.C. 20332
(M)

Soviet Studies
   (A Quarterly Journal of
   the USSR and Eastern Europe)
University of Glasgow
Glasgow G12 SQG, U.K.
(Q), NA

Soviet World Outlook
   (A Monthly Report on the View
   from the Kremlin on Issues
   Critical to U.S. Interest)
Advanced International Studies
   Institute, Inc.
4330 East-West Highway,
Suite 1122
Washington, D.C. 20014
(M)

Spettatore Internazionale
   (English Edition)
Instituto Affari Internazionali
Viale Mazzi 88
00195 Rome, Italy
(Q), 1965, NA

Standardization Newsletter
DOD, Defence Standardization
   Committee
Victoria Barracks
Melbourne, Victoria, 3004
Australia
(Q), 1969-

Strategic Analysis
Institute for Defense Studies
   and Analysis
Sapru House
Barakhamba Road
New Delhi, 11001, India
(M), 1977-, AMB

Strategic Digest
Institute for Defence Studies
   and Analysis
Sapru House
Barakhamba Road
New Dehli, 11001, India
(M), 1971-, AMB

Strategic Latin American Affairs
Copley & Associates, S.A.
2030 M Street, N.W., Suite 602
Washington, D.C. 20036
(W)

Strategic Review
United States Strategic Institute
1612 K Street, N.W., Suite 1204
Washington, D.C. 20006
(Q), AU, NA, AMB

Strategic Studies
Institute of Stategic Studies
P.O. Box 1173
Islamabad, Pakistan
(Q)

Strategy and Tactics
   (The Magazine of Conflict
   Simulation)
Simulations Publications, Inc.
44 East 23rd Street
New York, NY 10010
(BM), AMB

Strategy Week
Strategy Publications, Inc.
2030 M Street, N.W.
Washington, D.C. 20036
(W), 1980-

*Studies in Comparative
   Communism*
   (An International
   Interdisciplinary Journal)
Von Kleinsmid Institute of
   International Affairs
School of International Relations
University of Southern California
University Park
Los Angeles, CA 90007
(Q)

*Surface Warfare Magazine
DCNO (Surface Warfare)
OP-03AX
1300 Wilson Blvd., Rm. 782
Arlington, VA 22209
(M), NA

Survey
   (A Journal of East and West
   Studies)
Ilford House
133 Oxford Street
London, W1R 1TD, U.K.
(Q)

Survival
International Institute for
   Strategic Studies
18 Adam Street
London, WC2N 6AL, U.K.
(BM), 1959-, NA, AMB

Swiss Review of World Affairs
   (A Monthly Publication of the
   Neue Zurcher Zeitung)
11 Falkenstrasse
8001 Zurich, Switzerland
(M)

Talon
Box 6066
USAF Academy, CO 80840
(M/10)

Tank
Royal Tank Regiment Pubs., Ltd.
1 Elverton Street
London, SW1P 2QJ, U.K.
(Q), 1919-

TAVR Magazine
   (Journal of the Territorial
   & Army Volunteer Reserve)
Council of the TA and VRS
Centre Block
Duke of York HQ
Chelsea, London, SW3 4SG, U.K.
(M)

TIG Brief
HQ, USAF/IGEP
Washington, D.C. 20330
(BW), AU

To the Point
P.O. Box 78440
Sandton 2146, RSA
(W)

*Translog
   (The Journal of Military
   Transportation Management)
MTMC
Washington, D.C. 20315
(M), AU

Transportation Brief
HQ, USAF/LET
Washington, D.C. 20330

Underwater Letter
Callahan Publications
6631 Old Dominion Drive
McLean, VA 22101
(BW)

USI Journal
   (India's Oldest Journal on
   Defence Affairs)
United Service Institute
Kashmir House
King George's Avenue
New Dehli, 110011, India
(Q), 1870-, AMB

USAF Fighter Weapons Review
57th TTW/ DON (TAC)
Nellis AFB, NV 89191
(M)

*United States Army Aviation
    Digest
Drawer P, Attn: ATZQ-TD-AD
Ft. Rucker, AL 36362
(M), AU, AMB

U.S. Army Recruiting and
    Reenlistment Journal
USARC
Ft. Sheridan, IL 60037
(M), 1919-

USNI Proceedings
U.S. Naval Institute
Annapolis, MD 21402
(M), 1917-, AU, NA, AMB

*U.S. Navy Medicine
BUMED (Code 0010)
Washington, D.C. 20372
(M)

Vantage Point
    (Developments in North Korea)
Vantage Point
Naewoe Press
194 Wonnah-Dong, Chongno-Gu
Seoul 110, Korea
(M), 1978-

Vikrant
    (Asia's Defence Journal)
1 Todarmal Road
Bengali Market
New Dehli, 11001, India
(M)

Waggoner
Royal Corps of Transport
Regimental HQ
Buller Barracks
Aldershot, U.K.
(Q), 1891-

War/Peace Report
Center for War/Peace Studies
218 E. 18th Street
New York, NY 10003

Warship
Conway Maritime Press, Ltd.
2 Nelson Road
Greenwich, London SE10, U.K.
(Q), 1977-

Washington Notes on Africa
Washington Office on Africa
110 Maryland Avenue, N.E.
Washington, D.C. 20002
(Q)

The Washington Quarterly
    (A Review of Strategic
    and International Issues)
CSIS, Georgetown University
1800 K Street, N.W.
Washington, D.C. 20006
(Q), NA

Western European Politics
Frank Cass & Co., Ltd.
Gainsborough House
Gainsborough Road
London E111RS, U.K.
(Q)

Wings of Gold
Association of Naval Aviation,
    Inc.
P.O. Box 4821
Pensacola, FL 32507
(Q)

World Affairs
    (A Quarterly Review of
    International Problems)
American Peace Society
4000 Albermarle St., N.W.
Room 504
Washington, D.C. 20016
(Q), 1837-, NA

World Politics
    (A Quarterly Journal
    of International Relations)
Center for International Studies
Princeton University
Corwin Hall
Princeton, NJ 08540
(Q), 1948-, NA, AMB

The World Today
   (Under the Auspices of the
   Royal Institute of Int'l Affairs)
Chatham House
10 St. James Square
London SW1Y 4LE, U.K.
(M), 1945-, NA

Worldview
Council on Relations and
   International Affairs
170 East 64th Street
New York, NY 10021
(M)

Yugoslav Survey
   (A Record of Facts and
   Information)
Jugoslovenski Pregled
Mose Pijade 8-1
Belgrade, Yugoslavia
(Q), 1960-

Zosen (Shipbuilding)
Tokyo News Service, Ltd.
Kosoku Doro Nishi
8-10 Ginza Nishi
Cho-ku, Tokyo 104, Japan
(M), 1956-, **NA**

APPENDIX B: ORGANIZATIONS

There are literally hundreds of organizations which conduct research in the military and strategic fields. They range from independent non-profit organizations to university centers to defense industry associations. They publish journals, magazines and newsletters; have libraries and clipping files; and have issue experts who can be of great assistance to journalists, researchers, and students. The 60 organizations listed in this Appendix are the major organizations in each category. Their major publications are also listed. Most of the organizations have libraries that are open to qualified researchers.

REFERENCE WORKS:

A number of standard reference works found in almost any library which list and describe organizations are:

The Encyclopedia of Associations (Detroit, MI: Gale Research Co., 3 vols., 1975-)
New Research Centers (Detroit, MI: Gale Research Co., 1968)
The Middle East and North Africa
Africa South of the Sahara
Far East and Australasia

Two other reference works which describe worldwide organizations in the military and strategic fields are:

National Security Affairs: A Guide to Information Sources
Survey of Strategic Studies (London, IISS (Adelphi Papers #64), 1970)

Reference works on the organizations in Washington include:

Washington Representatives, 19__ (4th Ed., Washington, D.C.: Columbia Books, Inc., 1976-)
Scholar's Guide to Washington, D.C. for . . . (Washington, D.C.: Smithsonian Institution Press)
    Latin American and Caribbean Studies
    Russian/Soviet Studies
    East Asian Studies
    African Studies

Washington V: A Comprehensive Directory of the Nation's Capitol . . . It's People and Institutions (Carl T. Grayson, Jr., ed., Washington, D.C.: Potomac Books, Inc. 1979)

Washington Information Directory (Washington, D.C.: Congressional Quarterly, Inc.)
Library and Reference Facilities in the Area of the District of Columbia (9th Ed., Washington, D.C.: American Society for Information Science)

ORGANIZATIONS:

Advanced International Studies
  Institute
4330 East-West Highway,
  Suite 1122
Washington, D.C. 20014
(301) 951-0818
  *Soviet World Outlook,
  Monographs in International
  Affairs Series*

The Africa Fund
  (Associated with the American
  Committee on Africa)
198 Broadway
New York, NY 10038

African Bibliographic Center
1346 Connecticut Ave., N.W.
Washington, D.C. 20036
(202) 223-1392
  *A Current Bibliography on
  African Affairs*

Air Force Association
1750 Pennsylvania Ave., N.W.
Washington, D.C. 20006
(202) 637-3300
  Air Force Magazine

American Committee on Africa
198 Broadway
New York, NY 10038
(212) 962-1210
  ACOA Action News

American Defense Preparedness
  Association
Rosslyn Center
1700 N. Moore St., Suite 900
Arlington, VA 22209
(703) 522-1820
  National Defense

American Enterprise Institute
  for Public Policy Research
Foreign and Defense Policy
  Studies
1150 17th St., N.W.
Washington, D.C. 20036
(202) 862-5800
  *AEI Foreign Policy and
  Defense Review*

American Friends Service
  Committee (AFSC)
1501 Cherry St.
Philadelphia, PA 19602
(215) 241-7177
  Quaker Service Bulletin

American Security Council
499 S. Capitol St., S.W., Suite 500
Washington, D.C. 20003
(202) 484-1677
  *Washington Report,
  International Security Review*

Armament & Disarmament
  Information Unit (ADIU)
Mantell Building, University of
  Sussex
Falmer, Brighton BN1 9RF, U.K.
0273-686758
  *ADIU Report
  ADIU Occasional Papers*

Arms Control Association
11 Dupont Circle, N.W., Suite 900
Washington, DC 20036
(202) 797-6450
  Arms Control Today

Association of the U.S. Army
2425 Wilson Blvd.
Arlington, VA 22201
(703) 841-4300
  Army, Special Reports

229

Atlantic Council of the United
   States
1616 H St., N.W.
Washington, D.C. 20006
(202) 347-9353
   *Atlantic Community News/
   Quarterly, Atlantic Council
   Security Series Policy Papers*

The Brookings Institution
Foreign Policy Studies Program
1775 Massachusetts Ave., N.W.
Washington, D.C. 20036
(202) 797-6000
   Studies in Defense Policy

Campaign Against the Arms
   Trade
5 Caledonian Road
London N1 9DX, U.K.
01-2781976
   CAAT Newsletter

Center for Defense Information
122 Maryland Avenue, N.E.
Washington, D.C. 20002
(202) 543-0400
   Defense Monitor

Center for International Affairs
Harvard University
6 Divinity Avenue
Cambridge, MA 02138
(617) 495-4420
   *Harvard Studies in
   International Affairs*

Center for International and
   Strategic Affairs
University of California,
   Los Angeles
405 Hilgard Avenue (JG-35)
Los Angeles, CA 90024
   *Arms Control and International
   Security Working Papers*

Center for International Policy
120 Maryland Ave., N.E.
Washington, D.C. 20002
(202) 544-4666
   International Policy Report

Center for National Security
   Studies
122 Maryland Ave., N.E.
Washington, D.C. 20002
(202) 544-5380
   First Principles

Center for Naval Analysis
2000 N. Beauregard St.
Alexandria, VA 22311
(703) 998-3500
   Naval Abstracts

Center for Strategic and
   International Studies
Georgetown University
1800 K St., N.W., Suite 400
Washington, D.C. 20006
(202) 887-0200
   *Washington Review of Strategic
   and International Studies,
   Allied Interdependence
   Newsletter, Washington Papers*

Coalition for a New Foreign
   and Military Policy
120 Maryland Ave., N.E.
Washington, D.C. 20002
(202) 546-8400
   Close Up

Coalition for Peace Through
   Strength
American Security Council
Boston, VA 22713
(703) 825-1776

Committee on the Present Danger
1800 Massachusetts Ave., N.W.
   Suite 601
Washington, D.C. 20036
(202) 466-7444

Council on Economic Priorities
84 Fifth Avenue
New York, NY 10011
(212) 691-8550
   *CEP Studies/Reports/
   Newsletters, CIC Update*

Federation of American
 Scientists
307 Masschusetts Ave., N.E.
Washington, D.C. 20002
(202) 546-3300
 *FAS Newsletter/Public
 Interest Report*

Foreign Affairs Research
 Institute
Arrow House
27-31 Whitehall
London, 5W1A 2BX, U.K.
 FARI Papers

Foreign Policy Research Institute
University of Pennsylvania
133 S. 36th St.
Philadelphia, PA 19102
 FPRI Monographs, Orbis

Friends Committee on National
 Legislation (FCNL)
245 Second St., N.E.
Washington, D.C. 20002
(202) 547-4343
 Washington Newsletter

Heritage Foundation
513 C St., N.E.
Washington, D.C. 20002
(202) 546-4400
 *National Security Record,
 Policy Review, Backgrounder*

Hessische Stiftung Friedens
 und Konflictforschung (HSFK)
Leimenrode 29 6
6000 Frankfurt/Main 1, FRG

Hudson Institute
Quaker Ridge Road
Croton-on-Hudson, NY 10520
(914) 762-0700

IFSH-Working Group on Arms
 and Development
Von Melle Park 15
2000 Hamburg 13, FRG

Institute for Defense and
 Disarmament Studies
251 Harvard St.
Brookline, MA 02146
(617) 734-4216

Institute for Foreign Policy
 Analysis, Inc.
Central Plaza Bldg., 10th Floor
675 Massachusetts Ave.
Cambridge, MA 02139
 *Foreign Policy Reports, Special
 Reports*

Institute for Policy Studies/
 Transnational Institute
1901 Q St., N.W.
Washington, D.C. 20009
(202) 234-9382
 IPS Issue Papers, Books

Institute for World Order
777 UN Plaza
New York, NY 10017
 Alternatives, Working Papers

International Confederation
 for Disarmament and Peace
6 Endsleigh Street
London WC1, U.K.
 ICDP Peace Press

International Institute for
 Strategic Studies
23 Tavistock Street
London WC2E 7NQ, U.K.
 Survival, Adelphi Papers

Marine Corps Association
P.O. Box 1775, MCB Quantico
Quantico, VA 22134
 *Leatherneck, Marine Corps
 Gazette*

Members of Congress for Peace
 through Law
3538 House Annex 2
Washington, D.C. 20515
(202) 225-8550
 MCPL Legislative Alert

National Action/Research on
  the Military Industrial Complex
  (NARMIC)
American Friends Service
  Committee
1501 Cherry St.
Philadelphia, PA 19102
(215) 241-7175

National Strategy Information
  Center, Inc.
111 E. 58th St.
New York, NY 10022
(212) 838-2912
  *Agenda Papers, Strategy
  Papers*

Navy League of the United States
818 18th St., N.W.
Washington, D.C. 20006
(202) 298-9282
  Seapower

North American Congress on
  Latin America
151 West 19th Street, 9th Floor
New York, NY 10011
(212) 989-8890
  NACLA Report on the Americas

Nuclear Weapons Facilities
  Project (AFSC)
1428 Lafayette St.
Denver, CO 80218
(303) 832-4508

Pacific-Asia Resource Center
P.O. Box 5250
Tokyo International, Japan
  AMPO

Peace Research Institute of Oslo
Radhusgatan 4
Oslo 1, Norway
  *Bulletin of Peace Proposals,
  Journal of Peace Research*

Polemologisch Instituut,
  Groningen
Rijksstraatweg 76
9752 Haren, Netherlands

Rand Corporation
1700 Main Street
Santa Monica, CA 90401
  *Selected Rand Abstracts, Rand
  Papers, Reports, Studies*

SANE (A Citizen's Organization
  for a SANE World)
514 C St., N.E.
Washington, D.C. 20002
(202) 546-7100
  *SANE World, the Conversion
  Planner*

Stockholm International Peace
  Research Institute (SIPRI)
Bergshamra, S171 73 Solna
Sweden
  *SIPRI Yearbook, Books,
  Papers*

Strategic and Defence Studies
  Center
Australia National University
P.O. Box 4
Canberra, ACT 2600, Australia
  *SDSC Working Papers,
  Canberra Papers*

Tampere Peace Research Institute
Hameekatu 13 b A
P.O. Box 447
SF-33101 Tampere-19, Finland
  *Current Research on Peace
  and Violence*

U.S. Naval Institute
Annapolis, MD 21402
  USNI Proceedings

Washington Office on Africa
110 Maryland Ave., N.E.
Washington, D.C. 20002
(202) 546-7961
  Washington Notes on Africa

Washington Office on Latin
  America
110 Maryland Ave., N.E.
Washington, D.C. 20002
(202) 544-8045
  Update

IPS PUBLICATIONS

Real Security: Restoring American Power in a Dangerous Decade
By Richard J. Barnet

"*Real Security* is a *tour de force*, a gift to the country. One of the most impassioned and effective arguments for sanity and survival that I have ever read."—Dr. Robert L. Heilbroner

"An inspired and inspiring achievement... a first salvo in the campaign to turn our current security policies—diplomatic, military, and economic—in the direction of rationality. It may well be the basic statement around which opponents of unalloyed confrontation can gather. It will have great impact."—John Marshall Lee, Vice Admiral, USN (Ret.)

"As a summary of the critical literature on the arms race, Barnet's brief essay is an important antidote to hawkish despair."—*Kirkus Reviews*

$10.95 (paper, $4.95).

The Counterforce Syndrome: A Guide to U.S. Nuclear Weapons and Strategic Doctrine
By Robert C. Aldridge

This study discloses the shift from "deterrence" to "counterforce" in U.S. strategic doctrine. A thorough, newly-revised summary and analysis of U.S. strategic nuclear weapons and military policy including descriptions of MIRVs, MARVs, Trident systems, cruise missiles, and M-X missiles in relation to the aims of a U.S. first-strike attack. $4.95.

Dubious Specter: A Skeptical Look at the Soviet Nuclear Threat
By Fred Kaplan

Do the Soviets really threaten American ICBMs with a devastating surprise attack? Will Soviet military doctrine lead the Russians to threaten nuclear war in order to wring concessions from the West? Do Soviet leaders think they can fight and win a nuclear war? Fred Kaplan separates the myths from the realities about U.S. and Soviet nuclear stockpiles and strategies and provides the necessary background for understanding current debates on arms limitations and military costs. $4.95.

The New Generation of Nuclear Weapons
By Stephen Daggett

An updated summary of strategic weapons, including American and Soviet nuclear hardware. These precarious new technologies may provoke startling shifts in strategic policy, leading planners to consider fighting "limited nuclear wars" or developing a preemptive first strike capability. $3.00.

The Rise and Fall of the 'Soviet Threat': Domestic Sources of the Cold War Consensus
Alan Wolfe

A timely essay demonstrating that American fear of the Soviet Union tends to fluctuate according to domestic factors as well as in relation to the military and foreign policies of the USSR. Wolfe contends that recurring features of American domestic politics periodically coalesce to spur anti-Soviet sentiment, contributing to increased tensions and dangerous confrontations.

"At this moment, one could hardly want a more relevant book."—*Kirkus Reviews*. $4.95.

The Giants
Russia and America
By Richard Barnet

An authoritative, comprehensive account of the latest stage of the complex U.S.-Soviet relationship; how it came about, what has changed, and where it is headed.

"A thoughtful and balanced account of American-Soviet relations. Barnet goes beyond current controversies to discuss the underlying challenges of a relationship that is crucial to world order."—Cyril E. Black, Director, Center for International Studies, Princeton University. $4.95.

Toward World Security:
A Program for Disarmament
By Earl C. Ravenal

This proposal argues that new strategic weapons systems and increasing regional conflicts should prompt the U.S. to take independent steps toward disarmament including nondeployment of counterforce weapons. $2.00.

Supplying Repression: U.S. Support for Authoritarian Regimes Abroad
By Michael T. Klare and Cynthia Arnson

A comprehensive discussion of the programs and policies through which the U.S. supports police and internal security forces in repressive Third World countries.

"Very important, fully documented indictment of U.S. role in supplying rightist Third World governments with the weaponry and know-how of repression."—*The Nation*. $9.95 (paper, $4.95).

Beyond the Vietnam Syndrome: U.S. Interventionism in the 1980s
By Michael T. Klare

A study of the emergence of a new U.S. interventionist military policy. Shows how policymakers united to combat the "Vietnam Syndrome"—the public's resistance to American military involvement in future Third World conflicts—and to relegitimate the use of military force as an instrument of foreign policy. Includes a close look at the Pentagon's "Rapid Deployment Force," and a study of comparative U.S. Soviet transcontinental intervention capabilities. $4.95.

Resurgent Militarism
By Michael T. Klare and the Bay Area Chapter of the Inter-University Committee

An analysis of the origins and consequences of the growing militaristic fervor which is spreading from Washington across the nation. The study examines America's changing strategic position since Vietnam and the political and economic forces which underlie the new upsurge in militarism. $2.00.

Soviet Policy in the Arc of Crisis
By Fred Halliday

The crescent of nations extending from Ethiopia through the Arab world to Iran and Afghanistan has become the setting of an intense new geopolitical drama. In this incisive study, Halliday reviews the complex role played there by the Soviet Union—a role shaped as much by caution as by opportunity, as much by reaction to American moves as by Soviet initiative. Above all, the Soviet role is defined and limited by the indigenous politics of the region. $4.95.

After the Shah
By Fred Halliday

Important background information on the National Front, the Tudeh Party, the religious opposition and many other groups whose policies and programs will determine Iran's future. $2.00.

The Lean Years
Politics in the Age of Scarcity
By Richard J. Barnet

A lucid and startling analysis of basic global resources: energy, non-fuel minerals, food, water, and human labor. The depletion and maldistribution of supplies bodes a new global economic, political and military order in the 1980s.

"... brilliantly informed book ... cogent, aphoristic pulling together of the skeins of catastrophic scarcity in 'the coming postpetroleum world...'"—*Publishers Weekly*. $12.95.

Feeding the Few:
Corporate Control of Food
By Susan George

The author of *How the Other Half Dies* has extended her critique of the world food system which is geared towards profit not people. This study draws the links between the hungry at home and those abroad exposing the economic and political forces pushing us towards a unified global food system. $4.95.

Global Reach:
The Power of the Multinational Corporations
By Richard Barnet and Ronald Müller

"A searching, provocative inquiry into global corporations... Barnet and Müller are trenchant and telling in their discussion of the possible end of the nation-state, and have some penetrating views on 'economic imperialism' and future changes in employment patterns and the standard of living under the domination of the global oligopolists."—*Publishers Weekly* $7.95.

The Crisis of the Corporation
By Richard Barnet

Now a classic, this essay analyzes the power of the multinational corporations which dominate the U.S. economy, showing how the growth of multinationals inevitably results in an extreme concentration of economic and political power in a few hands. The result, according to Barnet, is a crisis for democracy itself. $1.50.

Decoding Corporate Camouflage: U.S. Business Support for Apartheid
By Elizabeth Schmidt
Foreword by Congressman Ronald Dellums

By exposing the decisive role of U.S. corporations in sustaining apartheid, this study places highly-touted employment "reforms" in the context of the systematic economic exploitation and political repression of the black South African majority.

"... forcefully presented."—*Kirkus Reviews*. $4.95.

South Africa: Foreign Investment and Apartheid
By Lawrence Litvak, Robert DeGrasse, and Kathleen McTigue

A critical examination of the argument that multinationals and foreign investment operate as a force for progressive change in South Africa. "Its concise and well-documented debunking of the myth that foreign investment will eventually change the system of exploitation and repression in South Africa deserves wide readership... Highly recommended."—*Library Journal*. $3.95.

A Continent Beseiged: Foreign Military Activities in Africa Since 1975
By Daniel Volman

A study of the growing military involvement of the two superpowers and their allies in Africa. $2.00.

The Nicaraguan Revolution: A Personal Report
By Richard R. Fagen

Tracing the history of the Nicaraguan Revolution, Fagen focuses on six legacies that define current Nicaraguan reality: armed struggle; internationalization of the conflict; national unity; democratic visions; death, destruction and debts; and political bankruptcy. This primer on the state of Nicaraguan politics and economics provides an insightful view of the Sandinist quest for power and hegemony. The report contains twenty photographs by Marcelo Montecino and appendices with the basic documents necessary for understanding contemporary Nicaraguan affairs. $4.00.

Chile: Economic 'Freedom' and Political Repression
By Orlando Letelier

A trenchant analysis by the former leading official of the Allende government who was assassinated by the Pinochet junta. This essay demonstrates the necessary relationship between an economic development model which benefits only the wealthy few and the political terror which has reigned in Chile since the overthrow of the Allende regime. $1.00.

Assassination on Embassy Row
By John Dinges and Saul Landau

A devastating political document that probes all aspects of the Letelier-Moffitt assassinations, interweaving the investigations of the murder by the FBI and the Institute.

"... An engrossing study of international politics and subversion ..."—*Kirkus Reviews*. $14.95.

Human Rights and Vital Needs
By Peter Weiss

Delivered one year after the assassination of Orlando Letelier and Ronni Karpen Moffitt, this extraordinary address commemorates them by calling for a human rights policy that includes not only political and civil rights, but economic, social, and cultural rights as well. $1.00.

The Federal Budget and Social Reconstruction
Marcus Raskin, Editor

This study describes the Federal Budget, sets new priorities for government spending and presents alternative policies for defense, energy, health and taxation.

"The issuance of this report is a major political event and a challenge to mainstream ideology. It should be widely purchased."—*Library Journal*. $8.95.

Postage and Handling:
All orders must be prepaid. For delivery within the USA, please add 15% of order total. For delivery outside the USA, add 20%. Standard discounts available upon request.

Please write the Institute for Policy Studies, 1901 Que Street, N.W., Washington, D.C. 20009 for our complete catalog of publications and films.